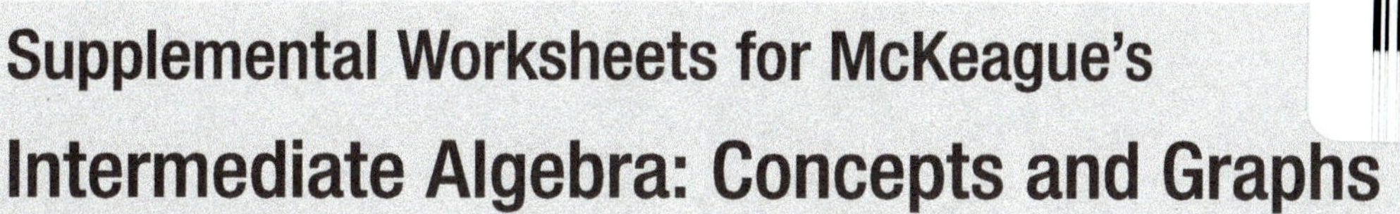

Supplemental Worksheets for McKeague's Intermediate Algebra: Concepts and Graphs

Matched Problems with Objectives

Prepared by

Judy Barclay

Supplemental Worksheets for McKeague's Intermediate Algebra: Concepts and Graphs Matched Problems with Objectives 2nd Edition

Publisher: XYZ Textbooks

Sales: Amy Jacobs, Richard Jones, Bruce Spears, Rachael Hillman

Cover Design: Rachel Hintz

Printed in the United States of America

ISBN-13: 978-1-936368-49-5 / ISBN-10: 1-936368-49-8

For product information and technology assistance, contact us at
XYZ Textbooks, 1-877-745-3499

For permission to use material from this text or product,
e-mail: **info@mathtv.com**

XYZ Textbooks
1339 Marsh Street
San Luis Obispo, CA 93401
USA

For your course and learning solutions, visit **www.xyztextbooks.com**

Learning Objectives

Chapter 1

1.1 Recognizing Patterns

Objective A Recognize a pattern in a sequence of numbers.

Recognizing patterns is a very important skill in mathematical problem solving. Here we practice this skill with recognizing patterns of numbers. Patterns lead us to mathematical models that are used to solve application problems.

Objective B Extend an arithmetic sequence.

When each number in a sequence is found by adding the same amount to the previous number, then we call the sequence an arithmetic sequence. The amount we add each time is called the common difference. With these sequences it is also true that if you subtract any two consecutive terms you get the same number each time, and that number is called the common difference.

Objective C Extend a geometric sequence.

A geometric sequence is a sequence of numbers in which each term comes from the previous term by multiplying by the same amount each time. The amount you multiply by each time is called the common ratio. Another way to say the same thing is to say that, if you divide any term by the previous term, you get the common ratio.

Objective D Recognize and extend a Fibonacci sequence.

The Fibonacci sequence is an example of a pattern that can be used to predict events we see in the world around us. In the example in the book, it can predict the number of bees in each generation of the family tree of a male honeybee. Try a search on the Internet for Fibonacci sequence and see what pops up.

Objective E Simplify expressions using the rule for order of operations.

Here we review one of the most important rules in this course. Be sure that you master it because we will use it many times throughout the course. It is a simple rule to follow. First we evaluate any numbers with exponents; then we multiply and divide; and finally we add and subtract, always working from left to right when more than one of the same operation symbol occurs in a problem.

Objective F Graph numbers on the real number line.

The numbers on the number line increase in size as we move to the right. When we compare the size of two numbers on the number line, the one on the left is always the smaller number.

Objective G List the elements in subsets of the real numbers.

The set of real numbers has several important subsets: natural numbers, whole numbers, integers, rational numbers, and irrational numbers. It is important to be able to recognize all of these subsets. Irrational numbers can be approximated using a calculator or table. Be sure not to mix up the exact value of a number and its approximation.

1.2 Products

Objective A Recognize and apply the associative property of multiplication.

Now we start our review of the properties of numbers. These properties are descriptions of the way in which our brains should work when dealing with numbers and operation symbols. The associative property of multiplication deals with the grouping of numbers. It tells us that the answer will be the same if we change the grouping of the numbers.

Objective B Recognize and apply the distributive property.

We say multiplication distributes over addition (and subtraction). Although the distributive property is stated as $a(b+c)$, it can be extended to include sums of more than two numbers. The important thing is that the numbers within the parentheses are added together, because multiplication distributes over addition, but it does not distribute over multiplication.

Objective C Simplify algebraic expressions by combining similar terms.

Similar terms are terms that have the same variable part. When combining similar terms, we are actually using the distributive property to factor out the common variable part.

Objective D Simplify algebraic expressions using the properties for exponents.

Here we review three very important properties of exponents. Memorize them, if you haven't already. We use the associative and distributive properties to simplify algebraic expressions.

Objective E Review multiplication with polynomials.

Polynomials are some of the simplest expressions to work with in algebra. Here we review how to multiply polynomials, a skill that you will need many times throughout this course and any math courses you take after this one.

Objective F Write an expression for revenue using algebraic expressions.

This is an example of an application that uses algebraic expressions. You will see it again in this course.

1.3 Sums and Differences

Objective A Recognize and use the commutative, associative, and distributive properties to simplify expressions.

An algebraic expression contains numbers, variables, operation symbols, and grouping symbols. Here we review how we simplify these expressions using several properties in the process. Then we extend the process to adding and subtracting polynomials.

Objective B Find the value of an algebraic expression.

This objective gives us a review of a technique we use often in algebra: finding the value of an expression for a specific value of a variable. An algebraic expression will take on different values depending on what value is assigned to each variable.

1.4 Factorizations

Objective A Factor positive integers into the product of primes.

It doesn't matter how you start factoring a number. When you have factored completely so that the only factors remaining are prime numbers, the result will always be the same. That is, if you start factoring 12 by writing it as 4 times 3, or if you start by writing it as 6 times 2, when you are done factoring, the answer will be the same in both cases.

Objective B Reduce fractions to lowest terms.

Factor the numerator and denominator into a product of primes first. Then look for common factors to divide out.

Objective C Factor an expression by grouping.

Factoring by grouping involves factoring the greatest common factor from the first two terms and the second two terms. Next we factor out a common factor again. If you do not have a common factor, you might have to rearrange the terms or check for mistakes in your factoring of the greatest common factor.

Objective D Factor a trinomial by trial and error.

You will factor many trinomials as you progress through this course. Here is your chance to review this type of factoring and gain confidence with factoring so that it is an asset for you in your journey through algebra.

Objective E Factor the difference of two squares.

This is a formula that you should memorize: $a^2 - b^2 = (a+b)(a-b)$. And remember, the sum of two squares is not factorable.

Objective F Factor the sum and difference of two cubes.
You should memorize these two formulas as well.

Objective G Review of factoring polynomials in general.
The first step in any factoring problem is to factor out the greatest common factor if it is other than 1. The greatest common factor is the largest monomial factor that divides each term of the polynomial. After factoring this out, refer to the steps outlined in Section 1.4, Example 5 of the textbook. It is extremely important that you master factoring at this point in the course because we use it in many different situations.

1.5 Quotients

Objective A Simplify quotients using the rule for order of operations.
Most of these problems should be review. However, that doesn't mean that they are easy. As we stated earlier, the rule for order of operations is extremely important to your success in this course, and that is why we are reviewing it often in this first chapter. If you are having problems with these exercises, make sure to get some help and master this skill.

Objective B Use the properties of exponents to simplify expressions containing negative integer exponents.
The most important idea to keep in mind here is that a negative sign in front of an exponent tells us we are to take the reciprocal of the base. Negative exponents do not give us negative numbers.

Objective C Find the value of expressions containing quotients.
Again, finding the value of an expression is a very important skill in this course, and that is why we are reviewing it often in this chapter.

1.6 Unit Analysis and Scientific Notation

Objective A Use the appropriate conversion factor to convert units.
The ability to convert an expression with one set of units, like miles/hour, into an equivalent expression in another set of units, like feet/second, is a very valuable skill. You will see it again in many different courses, including nursing, chemistry, and trigonometry, among others. If you plan to take any of these courses, you might want to mark this section to refer back to.

Objective B Use scientific notation to write numbers and to simplify expressions.
You will see scientific notation again in many other courses, especially science courses. And it is built in to most scientific calculators. Scientific notation allows us to work with very large and very small numbers that would be difficult to manage otherwise.

Learning Objectives

Chapter 2

2.1 Linear and Quadratic Equations

Objective A Solve a linear equation in one variable.
This is one of the most important skills in algebra. You will need it throughout this course. If your equation has fractions or decimals, use the multiplication property of equality to eliminate the fractions or decimals first. The resulting equation will be easier to solve than the original one.

Objective B Solve equations by factoring.
We solve the equations here differently than we solve linear equations. You want to be able to look at the equation and tell which method of solving it is appropriate. For the equations that require factoring, be sure to write the equation in standard form first; you need 0 on one side before you factor the other side. As you can see, you must be able to factor polynomials to solve these problems. Factoring will continue to be an important skill throughout this course. If you are having trouble factoring, get help now.

Objective C Recognize and solve identities and equations with no solution.
We usually have special cases with all the different types of equations we solve. So here is our first introduction to special cases. If you are solving an equation, and the variable drops out and you are left with a true statement, the original equation is an identity. If the variable drops out and you get a false statement, then the original equation has no solution.

2.2 Formulas

Objective A Solve a formula with numerical replacements for all but one of its variables.
Working with formulas will be an important skill throughout this course and for other math classes you may take in the future. It is also important in most science classes, as well as business, nursing, and many other classes as well.

Objective B Solve formulas for the indicated variable.
How do you know when you have finished solving a formula for one of its variables? You know that you are finished when that variable appears alone on one side of the equal sign and not on the other side.

2.3 Applications

Objective A Apply the Blueprint for Problem Solving to a variety of application problems.
Problem solving is one of the most important skills you will gain in this course. In this section you are given a list of steps to solve applications and then a variety of problems to solve. Look back at the examples if you are having trouble with any of the exercises. It takes a great deal of practice to be successful at solving applications. Don't get frustrated—keep working on these.

Objective B Use a formula to construct a table of paired data.
You can generate a table using a graphing calculator. Check with your instructor to see if you can use a graphing calculator in this course. If not, you can make the table by hand.

2.4 Linear Inequalities in One Variable

Objective A Solve a linear inequality in one variable and then graph the solution set.
Solving linear inequalities is very similar to solving linear equations with one important difference: When you multiply or divide both sides of the inequality by a negative number, it reverses the inequality symbol.

Objective B Solve a linear inequality in one variable and then write the solution set using interval notation.
Interval notation is a very convenient way to express intervals. It will be used from now on in this course and in other math courses you will take in the future.

Objective C Solve a compound and continued inequalities and then graph the solution set.
Compound and continued inequalities deal with the union and intersection of the sets. You must understand union and intersection before being able to understand these inequalities. Get help with union and intersection if you are having difficulty here.

Objective D Solve application problems using inequalities.
Learning to solve applications is one of the most important skills you will take from this course. In this section you will see a variety of applications involving inequalities. Keep practicing application problems. You will eventually succeed with them.

2.5 Equations with Absolute Value

Objective A Solve equations with absolute value symbols.
The key to solving equations with absolute value is to rewrite them as equations without absolute value. To do so we can think of absolute value as indicating distance from the origin on the number line. If a number is five units from the origin, it can be five units to the right of the origin, or it can be five units to the left of the origin. Therefore, most of the equations we solve with absolute value are written first as two equations without absolute value, which in turn give us two distinct solutions. However, there are times when you will not get any solution to an absolute value equation, and other times when you will get exactly one solution.

2.6 Inequalities Involving Absolute Value

Objective A Solve inequalities with absolute value and graph the solution set.
To solve inequalities that contain absolute value, we rewrite them as equivalent inequalities without absolute value. One way to do this is to think of absolute value as indicating a distance from 0 on the number line. The result is, when solving an absolute value inequality with a $<$ symbol, the solution set will be the **intersection** of two sets. When solving an absolute value inequality with a $>$ symbol, the solution set will be the **union** of two sets. As with all problems in algebra, the more of these problems you attempt, the better you will become at solving them, even if, at first, they seem confusing and difficult.

Learning Objectives

Chapter 3

3.1 Paired Data and the Rectangular Coordinate System

Objective A Graph ordered pairs on a rectangular coordinate system.

All points on the x-axis will have a y-coordinate of zero and all points on the y-axis will have an x-coordinate of zero.

Objective B Graph linear equations by making a table.

It takes only two points to draw a line. However, it is a good idea to find a third point to check. If the three points do not line up, then you know you have made a mistake.

Objective C Graph equations using vertical translations.

Vertical shifts or translations move a graph up if the constant is positive and down if the constant is negative. These shifts are very helpful in graphing many types of equations.

Objective D Graph equations using intercepts.

Graphing linear equations by finding the intercepts works best when the coefficients of x and y are factors of the constant term.

Objective E Graph horizontal and vertical lines.

All horizontal lines have equations that are missing the x-value and all vertical lines have equations that are missing the y-value. This is because on a horizontal line, x can be anything, and on a vertical line, y can be anything.

3.2 The Slope of a Line

Objective A Find the slope of a line from its graph.

To find the slope from a graph, you must first find two points on the line. Then going from one point to the other, you count the vertical change (or rise) and the horizontal change (or run). The slope will be the quotient of the rise over the run.

Objective B Find the slope of a line given two points on the line.

Make sure to be consistent when subtracting the x- and y- values. You must start with the same point in both the numerator and denominator. Also make sure you divide the change in the y-values by the change in the x-values.

Objective C Use a graph to find the average speed of an object.

This is an application of slope. You can find any average rate by computing the slope between two points on the line.

Objective D Find the slope of a line parallel or perpendicular to a given line.

Parallel lines have the same slope and perpendicular lines have slopes that are negative reciprocals of one another.

3.3 The Equation of a Line

Objective A Find the equation of a line given its slope and y-intercept.

It is important to memorize the **slope-intercept form** of a the equation of a line, $y = mx + b$, where m is the slope and b is the y-intercept. It will also be useful in the future.

Objective B Find the slope and y-intercept from the equation of a line.

First you put the given equation into $y = mx + b$ form by solving the equation for y. Then the coefficient of x is the slope and the constant is the y-intercept.

Objective C Find the equation of a line given its slope and a point on the line.

It is important to memorize the **point-slope form** of the equation of a line, $y - y_1 = m(x - x_1)$, where m is the slope and (x_1, y_1) is a point on the line. It will also be useful in the future.

Objective D Find the equation of a line given two points on the line.

Here you use both the slope formula and the point-slope form of the equation of a line. First, you find the slope between the two given points. Then you substitute one point and the slope into the point-slope form.

Objective E Find the equation of a line parallel or perpendicular to a given line and through a given point. Write answer in standard form.

Remember from our past discussion parallel lines have the same slope and perpendicular lines have slopes that are negative reciprocals of each other. Standard form of a linear equation is $ax + by = c$ where a, b, and c are integers.

3.4 Linear Inequalities in Two Variables

Objective A Graph linear inequalities in two variables.

First graph the boundary line. Then shade in the correct region after testing a point not on the boundary. Remember to graph the boundary with a broken line when the inequality is $<$ or $>$. The only time you use a solid line for the boundary is when the inequality is $\le$ or $\ge$.

3.5 Introduction to Functions

Objective A Construct a table or a graph from a function rule.

Here you are given the equation of a function and the domain, or x-values. To construct a table, you substitute the x-values into the equation to find the corresponding y-values. They you graph these ordered pairs on the graph.

Objective B Identify the domain and range of a function or relation.

The domain is the set of inputs (or x-values) and the range is the set of outputs (or y-values).

Objective C Determine whether a relation is also a function.

The vertical line test is a very good way to determine if the graph of a relation is a function. If there are no vertical lines that can be found that cross the graph in more than one place, then the graph is the graph of a function.

3.6 Function Notation

Objective A Use function notation to find the value of a function for a given value of the variable.

Function notation is a very useful way to represent functions. You can display more information with fewer symbols. This notation does **not** mean to multiply f times x.

3.7 Variation

Objective A Set up and solve problems with direct, inverse, or joint variation.

There are many examples of direct, inverse, and joint variation in the sciences and math. Some of these are shown in this section. You will find others in your studies of geometry, chemistry, and physics.

3.8 Algebra and Composition with Functions

Objective A Find the sum, difference, product, and quotient of functions.

Many students find this notation confusing. Make sure you understand that $(f+g)(x)$ means to add together the two functions. It does **not** mean to multiply the sum by x. Also $(f+g)(3)$ means to find $f(3)+g(3)$. Again, it has nothing to do with multiplication by 3.

Objective B Find the composition of functions.

The notation for composition of functions can be very confusing. $(f \circ g)(x)$ means the same as $f[g(x)]$. Make sure you understand the difference between $(fg)(x)$ and $(f \circ g)(x)$. The first means to multiply the two functions and the second is the composition of the functions.

Learning Objectives

Chapter 4

4.1 Systems of Linear Equations in Two Variables

Objective A Solve systems of linear equations in two variables by the addition method.
There are many application problems that are described by a system of equations in two variables. So, it is very important to be able to solve systems of equations. For systems of two linear equations, this is one of the easiest methods, but like any skill in mathematics, it takes lots of practice.

Objective B Solve systems of linear equations in two variables by the substitution method.
This is a second method for solving systems of two linear equations. As you will see, this method can also be used when solving non-linear systems.

Objective C Solve systems of linear equations in two variables by graphing.
This is a third method for solving systems of two linear equations. This method is not the most accurate method, but it is very visual. It allows you to see when the lines are parallel and when they coincide. In all other cases, you can see that the lines intersect in exactly one point.

4.2 Systems of Linear Equations in Three Variables

Objective A Solve systems of linear equations in three variables.
We extend the method of solving systems of linear equation we learned in the last section to systems with three variables. You must be very neat and organized to be successful in solving these systems. It is a good idea to number your equations, as is done in the textbook, so you don't lose track of where you are.

4.3 Applications

Objective A Solve application problems whose solutions are found through systems of linear equations.
What makes these applications different from the ones we have encountered previously is that they have two variables instead of one. That's why systems of equations are good models for these applications.

4.4 Systems of Linear Inequalities

Objective A Graph the solution to a system of linear inequalities in two variables.
Graphing is the only method we have to show the solution to a system of linear inequalities. Make sure to shade one inequality at a time so you don't get lost. Also remember to use a dotted line when the boundary line is not part of the solution set.

Learning Objectives

Chapter 5

5.1 Basic Properties and Reducing to Lowest Terms

Objective A Reduce rational expressions to lowest terms.

It is much easier to work with rational expressions when they are in lowest terms. So this is a necessary skill. Be sure to only divide like factors, not like terms. This is a very common error.

Objective B Find function values for rational functions.

This notation can be confusing. The expression, $r(5)$, means that you should find the value of the function when x is 5. It does not mean to multiply the function by 5.

Objective C Work with ratios.

When you work with ratios that have variables in the denominators, you always have to make sure that the denominator does not equal zero. We always have to exclude any values that make the denominator equal zero.

5.2 Multiplication and Division of Rational Expressions

Objective A Multiply and divide rational expressions.

When you multiply rational expressions, you must completely factor each numerator and denominator first. Also remember that you can only reduce like factors, not terms.

When you divide rational expressions, you must multiply the first fraction by the reciprocal of the second. In other words, you multiply by the reciprocal of the expression that follows the division symbol.

5.3 Addition and Subtraction of Rational Expressions

Objective A Add and subtract rational expressions with the same denominator.

The process for adding and subtracting rational expression is the same as adding and subtracting fractions. If you are having trouble with these, go back and practice working with simple fractions.

Objective B Add and subtract rational expressions with different denominators.

Before finding the LCD, be sure to factor each denominator completely. After you add or subtract the rational expressions, check to make sure your answer is in lowest terms.

5.4 Complex Fractions

Objective A Simplify complex fractions.

There are two important reasons for learning how to simplify complex fractions. Simple fractions are much easier to work with and much easier to understand. In your future work with algebra, always simplify complex fractions before working with them.

5.5 Equations With Rational Expressions

Objective A Solve equations containing rational expressions.

In solving equations containing rational expressions, do not forget the check your answer in the original equation. If your answer makes any denominator equal to zero, it is called an extraneous solution and there is no solution to the problem.

Objective B Solve formulas containing rational expressions.

It will be necessary to solve for one variable in terms of other variables in algebra. Since there are many formulas that contain rational expressions, we practice solving these in this section.

Objective C Graph rational functions.

When graphing rational functions, you must find the vertical asymptote first and draw it in with a dotted line. It is important to remember that the rational function will never cross the vertical asymptote. It will get very close to it, but never touch it.

5.6 Applications

Objective A Solve application problems using equations containing rational expressions.

Solving application problems is one of the most important skills you will learn in algebra. In this section, we set up the problems and then solve them using equations that contain rational expressions.

Objective B Graph rational functions that have vertical and horizontal asymptotes.

When graphing rational functions, you must find the vertical and horizontal asymptotes first and draw them in with dotted lines. It is important to remember that the rational function will never cross the vertical asymptote. It will get very close to it, but never touch it. However, the rational function can cross a horizontal asymptote, as you will see later in your study of algebra.

5.7 Division of Polynomials

Objective A Divide a polynomial by a monomial.

A monomial is a single term. When you divide a polynomial by a monomial, you simply divide each term of the polynomial by the monomial.

Objective B Divide a polynomial by a polynomial.

When dividing a polynomial by another polynomial, it is not like dividing by a monomial. You must treat it like long division of whole numbers. Make sure that both of the polynomials are in decreasing powers of the variable and if there are any powers that are missing, fill in using a coefficient of 0.

Learning Objectives

Chapter 6

6.1 Rational Exponents

Objective A Simplify radical expressions using the definition of roots.

Working problems with this objective gets you more familiar with the definitions for roots. Every positive number has two square roots: One positive and one negative. Remember also that if you are trying to find an even root of a negative number, no real number will work. Later when we expand the type of numbers we work with to include complex numbers, we will see how to deal with square roots of negative numbers.

Objective B Simplify expressions with rational exponents.

Rational exponents are just another way to specify roots and powers of numbers. The further you go in mathematics the more useful rational exponents will be. For now, we are just trying to get used to the definition for rational exponents, and get a more intuitive feel for expressions containing rational exponents.

6.2 Simplified Form for Radicals

Objective A Write radical expressions in simplified form.

We put radicals in simplified form so that they are easier to work with. Sometimes this form is not a simpler-looking expression.

Objective B Rationalize a denominator that contains a single term.

In rationalizing a denominator, we are trying to make the denominator a perfect square, a perfect cube, or a perfect fourth root. We accomplish this by multiplying the numerator and denominator of the fraction by the appropriate radical.

6.3 Addition and Subtraction of Radical Expressions

Objective A Add and subtract radicals.

The main point here is that you can only add and subtract similar radicals. To see if the terms in an expression have similar radical parts, you must put each term in simplified form for radicals.

Objective B Construct golden rectangles from squares.

Golden rectangles have been studied for over 2,000 years. They are used in many different disciplines. Once we have a basic understanding of square roots and simplified form for radicals, we can construct golden rectangles and look for interesting relationships among the different parts of these rectangles.

6.4 Multiplication and Division of Radical Expressions

Objective A Multiply expressions containing radicals.

We multiply radical expressions in much the same way that we multiply polynomials. The skills you learn here will be used again later in the course.

Objective B Rationalize a denominator containing two terms.

When we rationalize a denominator we are changing it from an expression containing a radical to a rational number. In other words, we want to get rid of any radicals in the denominator. At the same time, we don't want to change the value of the expression we are working with. For that reason we must multiply the numerator and denominator by the same expression. And we have to choose that expression so that the multiplication we do in the denominator results in a rational number. That is why we use the conjugate of the denominator. The more of these problems you work, the easier they become.

6.5 Equations Involving Radicals

Objective A Solve equations containing radicals.

We are expanding the kinds of equations we can solve to include equations containing radicals. When we solve these equations it is very important to remember to check your solutions because any time we square both sides of an equation, we have the possibility that we have introduced an extraneous solution. So far we have had two types of equations where you must check the solutions. They are rational equations and radical equations.

Objective B Graph simple square root and cube root equations in two variables.

We are expanding the type of equations we can graph. When we graph equations containing radicals, we get a visual picture of how the radicals behave. You will learn more about this type of function later in your study of algebra.

6.6 Complex Numbers

Objective A Simplify square roots of negative numbers.

Complex numbers allow us to expand our work with radicals to include square roots of negative numbers. They also allow us to solve equations that do not have real answers.

Objective B Simplify powers of i.

It is interesting to note that whenever we raise i to a power, we will always get $i, -1, -i,$ or 1.

Objective C Solve for unknown variables by equating real parts and equating imaginary parts of two complex numbers.

This method gives us a different way to solve equations by comparing the coefficients.

Objective D Add and subtract complex numbers.

Adding and subtracting complex numbers is very similar to adding and subtracting like terms.

Objective E Multiply complex numbers.

Multiplying complex numbers is very similar to multiplying binomials, except that you can reduce i^2 to -1.

Objective F Divide complex numbers.

Dividing complex numbers is like rationalizing the denominators of radical expressions.

Learning Objectives

Chapter 7

7.1 Completing the Square

Objective A Solve quadratic equations by taking the square root of both sides.
This method is handy if your equation is in the correct form. The most important thing to remember here is that you need to find both the positive and the negative square root of the right side of the equation when using this method.

Objective B Solve quadratic equations by completing the square.
This method allows us to solve all quadratic equations whether or not they are factorable. Completing the square is the essential tool we need to derive the quadratic formula, which we will do in the next section. The technique of completing the square is something that you will use again in this book when you find the equation of a circle.

Objective C Use quadratic equations to solve for missing parts of right triangles.

There are many applications we can solve using the Pythagorean Theorem, $a^2 + b^2 = c^2$. You will see many more if you continue your study of mathematics. In applications, we often ignore negative answers because they do not make sense when compared with the original words of the problem.

7.2 The Quadratic Formula

Objective A Solve quadratic equations by the quadratic formula.
The quadratic formula is a very powerful formula. It allows us to solve all quadratic equations, whether or not they are factorable. Make sure that you memorize this formula. You will need it in the future.

Objective B Solve application problems using quadratic equations.
There are many applications that require you to use quadratic equations in the solution process. If the equation is factorable, then that's the way to solve it. If it doesn't factor, or you cannot factor it, then you need to be able to use the quadratic formula. Again, you should memorize the formula.

7.3 Additional Items Involving Solutions to Equations

Objective A Find the number and kind of solutions to a quadratic equation by using the discriminant.
Sometimes we want to know only how many solutions an equation has, but not what they are exactly. For quadratic equations, the discriminant is the key to the number of solutions. Working with the discriminant also helps you become more familiar with the quadratic formula, and the effect different parts of the formula have on the solutions to the equation.

Objective B Find an unknown constant in a quadratic equation so that there is exactly one solution.
This is an exercise to improve your critical thinking skills. You are applying what you just learned to a different situation.

Objective C Find an equation from its solutions.
Here we are reversing the process. You are given the solutions to an equation and using them to find the equation that they came from. These problems should strengthen your problem solving skills.

7.4 More Equations

Objective A Solve equations that are reducible to a quadratic equation.

If we can classify new equations in terms of old equations, then we don't need new techniques to solve the new equations. Here we are looking at a variety of equations that can be thought of as having the form of a quadratic equation, so we can apply our techniques for quadratics to these new equations. Solving these equations gives you practice applying what you have learned about solving quadratic equations. It also gives you more experience using substitution.

Objective B Solve application problems using equations quadratic in form.

You are given more applications to solve, and something new for the golden ratio.

7.5 Graphing Parabolas

Objective A Graph a parabola.

The graph of a quadratic function is called a parabola. This graph has many important uses and applications, as you will see in your study of algebra. Make sure you get plenty of practice with graphing parabolas.

Objective B Solve application problems using information from a graph.

In this section we present a few examples of applications of parabolas. You can use the vertex of a parabola to find the maximum or minimum in many problems. If you go on to take a business calculus class, you will see more of these types of problems.

Objective C Find an equation from its graph.

These problems should reinforce the characteristics of the graph of a parabola. You are given the information about a parabola and asked to find its equation. Reversing the process reinforces the concepts learned.

7.6 Quadratic Inequalities

Objective A Solve quadratic inequalities and graph the solution set.

Again we see how important factoring is in algebra. The way we solve quadratic inequalities is useful for other types of problems if you continue with mathematics past this course. It is a very useful technique.

Objective B Solve a rational inequality and graph the solution set. (Problem Set exercises 23 – 34 are similar.)

Here we extend our reasoning skills from Objective A to include inequalities that involve rational expressions. As we do, we get more practice with addition and subtraction of rational expressions.

Learning Objectives

Chapter 8

8.1 Exponential Functions

Objective A Find function values for exponential functions.
Exponential functions are a very important class of functions. Many applications can be modeled by an exponential function. You will need to use a scientific or graphing calculator to calculate the answers to some of the problems in this section. If possible, get a calculator that you can become familiar with.

Objective B Graph exponential functions.
Graphing exponential functions allows us to see what is happening with the function values as x gets very large and as x gets very small. We can see that the graph approaches a horizontal asymptote as x gets very large in the decay model or as x gets very small in the growth model. We can also see what these two models have in common and how they differ.

Objective C Solve applications involving exponential growth and decay.
The natural exponential function, $y = e^x$, is introduced here for applications involving continuous growth or decay. This is a very important application that appears often in math and science applications.

8.2 The Inverse of a Function

Objective A Find the equation of the inverse of a function.
You can obtain the inverse of a function by exchanging the x and the y in its equation. In other words, if (a,b) belongs to the function *f*, then (b,a) belongs to its inverse. Be careful with the notation for the inverse of a function, $f^{-1}(x)$; it does not indicate the reciprocal of the function. The -1 in the inverse function notation is not an exponent.

Objective B Graph a function and its inverse.
The graph of a function and its inverse shows you how the two are related geometrically; that the graph of a function and its inverse are symmetric about the line. Also, you can see from the examples that only one-to-one functions have inverses that are functions. These are very important concepts to remember. If you go on to take more math classes, the relationship between a function and its inverse become more important.

8.3 Logarithms are Exponents

Objective A Convert between logarithmic form and exponential form.
Converting back and forth between the different forms is good practice. Doing so reinforces the definition of a logarithm and allows you to have a more intuitive feel for logarithms and how they are related to the exponential expressions you have worked with in the past. This skill will be very useful in solving exponential and logarithmic equations.

Objective B Use the definition of logarithms to solve simple logarithmic equations.
Solving these simple logarithmic equations will help you solve more complicated equations later on.

Objective C Sketch the graph of a logarithmic function
Graphing logarithmic functions allows us to see the relationship between the graphs of exponential and logarithmic functions. They are inverses of each other and are symmetric about the line $y = x$. You will also see that the logarithmic function has a vertical asymptote. As x approaches 0, the y-value gets either very large or very small.

Objective D Simplify expressions involving logarithms.
These problems are simply good practice. They allow you to become more familiar with the properties of logarithms, so that they reinforce how you should think when you see expressions containing logarithms.

Objective E Solve applications involving logarithmic equations.
One application of logarithms is in measuring the magnitude of an earthquake. Another is finding how long it takes for money to double in an account with compound interest. You will find many applications as you continue in your study of algebra.

8.4 Properties of Logarithms

Objective A Use the properties of logarithms to convert between expanded for and single logarithms.
The properties of logarithms are very similar to the properties of exponents because logarithms are exponents. You should memorize the properties and lots of practice using them will help with that. The more of these problems you work, the easier it will be to when to use the properties and when the properties do not apply.

Objective B Use the properties of logarithms to solve equations that contain logarithms.
Here you are asked to apply the properties of logarithms to solve equations. In the next section, you will apply the properties to other equations and applications involving the common logarithm and natural logarithm.

8.5 Common Logarithms and Natural Logarithms

Objective A Use a calculator to find common logarithms.
If possible, get a scientific or graphing calculator that you can become familiar with. That way, it will be easier for you to compute logarithms on tests and quizzes. Remember that when the base is not written, it is assumed to be 10, or a common logarithm.

Objective B Use a calculator to find a number given its common logarithm.
Here we are reversing the process. You are finding the inverse function.

Objective C Solve applications that involve logarithms.
Some of these applications are very interesting and you will find them in other courses you have taken or will take in the future.

Objective D Simplify expressions containing natural logarithms.
Natural logarithms have many applications in higher level math classes, and in science.

8.6 Exponential Equations and Change of Base

Objective A Solve exponential equations.
It is important to learn how to solve these equations because you will be using them to solve applications at the end of this section. Remember to always isolate the exponential on one side before taking the log of each side.

Objective B Use the change-of-base property.
Since your calculators only allow you to evaluate logarithms with base *e* and base 10, it is important to learn how to use the change-of-base property. With this property you can graph $y = \log_3 x$ on your calculator.

Objective C Solve application problems whose solutions are found by solving logarithmic or exponential equations.
Many real world problems are modeled using exponential equations. Exponential and logarithmic equations can be very useful in solving applications in Biology, Business, Finance, and Economics, as well as physical science and math.

Learning Objectives

Chapter 9

9.1 More about Sequences

Objective A Write the terms of a sequence, given the general term.

These are just substitution problems. You substitute each value of n into the formula for the general term to get the terms of the sequence. Also, the general term of a sequence is the same as the nth term of the sequence. The general term is the same as function notation. Instead of writing $f(n)$, you write a_n.

Objective B Write terms of a sequence given recursively.

Another way to describe a sequence is by defining the general term in terms of the previous term. In other words, you do something to the previous term to get the next term. The notation can be confusing here: a_{n-1} is the previous term and a_n is the term you are currently finding.

Objective C Find the general term for a sequence.

It is often more difficult to find the general term for a sequence. Here you will be looking for patterns. Don't be afraid to guess and then see if your general term works for all the terms you are given.

9.2 Series

Objective A Expand and simplify a series given by summation notation.

These are just addition problems. You must know how to read the information given with summation notation in order to expand it into a string of terms that are connected by addition. A series is the sum of the terms in a sequence. Summation notation is a very handy and compact way of expressing a sum. It is used in courses of statistics and calculus. You can also use a graphing calculator to find these sums.

Objective B Write a series using summation notation.

Here you are going in the opposite direction from Objective A. You are given the expanded form of a series and are looking to rewrite it with summation notation. To do so, you need to find the general term of the sequence. As was the case in the previous section, there is some guessing and trial and error involved. However, in the next two sections, you will be given some formulas to find the general terms of certain types of sequences and series.

9.3 Arithmetic Sequences

Objective A Identify an arithmetic sequence (or arithmetic progression) and find its common difference.

Any sequence in which each term comes from the previous term by adding the same amount each time is an arithmetic sequence. The amount that you add each time is the common difference. You will find some very interesting applications of arithmetic sequences in many fields. You can look at the applications at the end of this section to see a few common ones.

Objective B Find the general term and the sum of an arithmetic sequence (or arithmetic progression) using the formulas.

Finding the nth term and the sum of an arithmetic sequence is like solving a puzzle. You have to take all the things that you know and try to fit them together to find the formula for either the general term or the sum. You are given two formulas to help you solve the puzzle.

9.4 Geometric Sequences

Objective A Identify a geometric sequence (or geometric progression) and find its common ratio.

Any sequence in which each term comes from the previous term by multiplying by the same amount each time is a geometric sequence. The amount you multiply by each time is the common difference. You will find some very interesting applications of geometric sequences in many fields. You can look at the applications at the end of this section to see a few common ones.

Objective B Find the general term and the sum of a geometric sequence (or geometric progression) using the given formulas.

Finding the *n*th term and the sum of a geometric sequence is like solving a puzzle. You have to take all the things that you know and try to fit them together to find the formula for either the general term or the sum. You are given two formulas to help you solve the puzzle. You should memorize these two formulas.

Objective C Find the sum of an infinite geometric series.

It is very interesting to discover that some infinite geometric series have a finite sum. It is important to make sure that the common ratio is between -1 and 1 before using the formula. You should memorize this formula also. It is an easy one to use.

Learning Objectives

Chapter 10

10.1 The Circle

Objective A Use the distance formula.
The distance formula comes from the Pythagorean Theorem. If you look at the derivation of the distance formula, it will help you in memorizing it. It will also assist you in memorizing the equation of a circle, which is our next objective.

Objective B Write the equation of a circle, given its center and radius.
As we stated above, if you've memorized the distance formula, you just square each side of the equation to get the standard form of a circle. Remember that (h,k) is the center and r is the radius. A common mistake is to forget to take the square root of the right side of the equation to get the radius.

Objective C Find the center and radius of a circle from its equation, and then sketch the graph.
Many times you will need to complete the square in order to find the equation of a circle. You learned the technique of completing the square when we solved quadratic equations. You should review this skill because you will need it in this chapter to put conic sections into standard form.

10.2 Ellipses and Hyperbolas

Objective A Graph an ellipse with center at the origin.
When the equation is written in standard form, the x-intercepts are the positive and negative square roots of the number below x^2. The y-intercepts are the positive and negative square roots of the number below y^2.

Objective B Graph a hyperbola with center at the origin.
When the equation is written in standard form, the sides of the rectangle used to draw the asymptotes are the positive and negative square roots of the number below x^2. The top and bottom of the rectangle are the positive and negative square roots of the number below y^2. It is important to know which way to draw the hyperbola. The hyperbola with x^2 first intersects the x-axis. The hyperbola with y^2 first intersects the y-axis.

Objective C Graph an ellipse with center at (h,k).
To draw the graph of an ellipse that has been shifted, first put the equation in standard form. Next draw a new set of translated axes at the point (h,k). Now proceed to draw the ellipse the same way you did with the center at the origin, but instead using the point (h,k) as the center.

Objective D Graph a hyperbola with center at (h,k).
To draw the graph of a hyperbola that has been shifted, first put the equation in standard form. Next draw a new set of translated axes at the point (h,k). Now proceed to draw the rectangle and the hyperbola the same way you did with the center at the origin, but instead using the point (h,k) as the center.

10.3 Second Degree Inequalities and Nonlinear Systems

Objective A Graph second-degree inequalities.
You are graphing regions in the plane. Remember to use a broken line when there is a less than or greater than symbol. Also remember to shade the appropriate region after graphing the second-degree equation.

Objective B Solve systems of nonlinear equations.
When solving a system that contains a second-degree equation and a linear equation, you must use the substitution method. Solve for one variable in the linear equation and then substitute it into the second-degree equation. You can use either the substitution or the elimination method when solving a system containing two second-degree equations.

Objective C Graph the solution sets to systems of inequalities.
You are looking for the intersection of two or more regions in the plane. Remember to use a broken line when there is a less than or greater than symbol. Also remember to shade the appropriate region after graphing both of the equations. The region they have in common is the solution to the problem.

Matched Problems with Objectives Name ____________________

Recognizing Patterns

Section 1.1 Date ____________________

Directions Each problem below is similar to the example with the same number in your textbook. After reading through an example in your textbook, or watching one of the videos of that example on MathTV, try the matched problem to check your progress in this section. The shaded text is the learning objective associated with the matched problems that appear below the objective.

Objective A Recognize a pattern in a sequence of numbers. (Problem Set exercises 1 – 14 are similar.)

1. Use inductive reasoning to find the next term in each sequence:

a. 3, 8, 13, 18, . . .

b. 6, 12, 24, 48, . . .

c. 1, 8, 27, 64, . . .

Objective B Extend an arithmetic sequence. (Problem Set exercises 15 – 24 are similar.)

2. Each sequence shown here is an arithmetic sequence. Find the next two numbers in each sequence:

a. 4, 7, 10, . . .

b. 4, 2, 0, . . .

c. $\frac{1}{2}, 2, \frac{7}{2}, \ldots$

Objective C Extend a geometric sequence. (Problem Set exercises 25 – 36 are similar.)

3. Each sequence shown here is a geometric sequence. Find the next two numbers in each sequence:

a. 3, 12, 48, . . .

b. $7, -14, 28, \ldots$

c. $\frac{1}{3}, \frac{1}{9}, \frac{1}{27}, \ldots$

Objective D Recognize and extend a Fibonacci sequence. (Problem Set exercises 41 – 42 are similar.)

4. Find the number of bees in the eleventh generation of the family tree of a male honeybee on page 6 of the textbook.

Matched Problems with Objectives Name ____________________

Recognizing Patterns

Section 1.1 Date ____________________

Objective E Simplify expressions using the rule for order of operations. (Problem Set exercises 43 – 58 are similar.)

5. Simplify: $6+4(5+7)$

6. Simplify: $4\cdot 3^2-3\cdot 2^3$

7. Simplify: $16-(3\cdot 4^2-25)$

8. Simplify: $36-14\div 2+4$

Objective F Graph numbers on the real number line.

9. Locate the following numbers on the real number line: $-3.2, -0.5, \frac{3}{4}, \sqrt{3}$, and 2.1

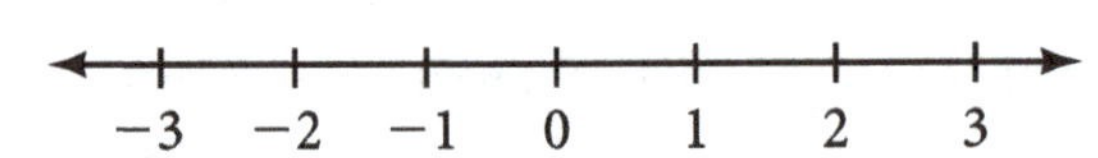

Objective G List the elements in subsets of the real numbers. (Problem Set exercises 59 – 66 are similar.)

10. For the following set, list the numbers that are (a) whole numbers, (b) integers, (c) rational numbers, (d) irrational numbers, and (e) real numbers: $\left\{-3, -2.1, \frac{1}{4}, \sqrt{2}, \sqrt{7}, 5\right\}$

Matched Problems with Objectives Name ______________________

Products

Section 1.2 Date ______________________

Directions Each problem below is similar to the example with the same number in your textbook. After reading through an example in your textbook, or watching one of the videos of that example on MathTV, try the matched problem to check your progress in this section. The shaded text is the learning objective associated with the matched problems that appear below the objective.

Objective A Recognize and apply the associative property of multiplication. (Problem Set exercises 1 – 14 are similar.)

1. Simplify: $7(6x)$

2. Simplify: $\frac{1}{8}(8x)$

3. Simplify: $16\left(\frac{3}{4}x\right)$

4. Simplify: $8\left(\frac{t}{2}\right)$

5. Simplify: $x\left(\frac{5}{x}\right)$

Objective B Recognize and apply the distributive property. (Problem Set exercises 15 – 60 are similar.)

6. Apply the distributive property to $3(5x+4)$.

7. Apply the distributive property to $\frac{1}{4}(8x+4)$.

Matched Problems with Objectives Name ____________________

Products

Section 1.2 Date ____________________

SECTION 1.2

8. Apply the distributive property to $0.05(x+400)$.

9. Multiply: $x\left(4+\frac{3}{x}\right)$

10. Simplify: $-9(3x-5)$

11. Simplify: $-(7a-14)$

Objective C Simplify algebraic expressions by combining similar terms.

12. Combine similar terms: $8x+3x$

Objective D Simplify algebraic expressions using the properties for exponents and the distributive and associative properties when necessary. (Problem Set exercises 61 – 66 are similar.)

13. Multiply: $(-4x^3)(6x^4)$

14. Simplify: $(-6x^3)^3(2x^8)$

Matched Problems with Objectives Name ______________________

Products

Section 1.2

Date ______________________

Objective E Review multiplication with polynomials (Problem Set exercises 67 – 100 are similar.)

15. Multiply: $5x^2\left(4x^2+8x+10\right)$

16. Multiply: $(2x-7)(3x-2)$

17. Multiply: $(3x-2)(4x+3)$

18. Multiply: $(2x-3y)\left(3x^2-2xy+y^2\right)$

19. Multiply: $(5x-2)^2$

Objective F Write an expression for revenue using algebraic expressions. (Problem Set exercises 101 – 102 are similar.)

20. A store selling memory sticks for home computers knows from past experience that it can sell x memory sticks each day at a price of p dollars per memory stick, according to the equation $x = 600 - 100p$. Write a formula for the daily revenue that involves only the variables R and p.

Matched Problems with Objectives Name ____________________

Sums and Differences

Section 1.3 Date ____________________

Directions Each problem below is similar to the example with the same number in your textbook. After reading through an example in your textbook, or watching one of the videos of that example on MathTV, try the matched problem to check your progress in this section. The shaded text is the learning objective associated with the matched problems that appear below the objective.

Objective A Recognize and use the commutative, associative, and distributive properties to simplify expressions. (Problem Set exercises 1 – 52 are similar.)

1. Simplify: $6x+5+3x+8$

2. Simplify the following by applying the distributive property: $5(4y+6)+10$

3. Simplify: $7+2(4y+8)+5y$

4. Simplify: $7-2(3x-5)+8x$

5. Simplify: $(x+9)(x-5)+4$

6. Add: $(4x^2-5x+9)+(2x^2+8x-7)$

7. Subtract: $(6x^2-5x+4)-(3x^2+4x-10)$

Matched Problems with Objectives Name ______________

Sums and Differences

Section 1.3 Date ______________

Objective B Find the value of an algebraic expression. (Problem Set exercises 53 – 64 are similar.)

8. Evaluate the expressions a^2+9 and $(a+3)^2$ for the following values of a:

a. $a=-1$

b. $a=0$

c. $a=2$

9. Evaluate the expression $-\frac{3}{4}x-2$ for the following values of x:

a. $x=0$

b. $x=4$

c. $x=-\frac{8}{3}$

10. Find the value of $3x-8y$ when

a. $x=-1$ and $y=3$

b. $x=2$ and $y=-4$

c. $x=3$

Matched Problems with Objectives Name ____________________

Factorizations

Section 1.4 Date ____________________

Directions Each problem below is similar to the example with the same number in your textbook. After reading through an example in your textbook, or watching one of the videos of that example on MathTV, try the matched problem to check your progress in this section. The shaded text is the learning objective associated with the matched problems that appear below the objective.

Objective A Factor positive integers into the product of primes. (Problem Set exercises 1 – 8 are similar.)

1. Factor 72 into a product of prime numbers.

Objective B Reduce fractions to lowest terms. (Problem Set exercises 9 – 12 are similar.)

2. Reduce $\frac{126}{210}$ to lowest terms.

Objective C Factor an expression by grouping. (Problem Set exercises 51 – 58 are similar.)

3. Factor: $2ax + 4bx + ay + 2by$

Objective D Factor a trinomial by trial and error. (Problem Set exercises 13 – 18 are similar.)

4. Factor: $6x^2 + 11x + 4$

Objective E Factor the difference of two squares. (Problem Set exercises 19 – 30 are similar.)

5. Factor: $9x^2 - 49$

Matched Problems with Objectives Name ____________________

Factorizations

Section 1.4 Date ____________________

SECTION 1.4

Objective F Factor the sum and difference of two cubes. (Problem Set exercises 31 – 38 are similar.)

6. Verify the two formulas:

$$a^3 + b^3 = (a+b)(a^2 - ab + b^2)$$

$$a^3 - b^3 = (a-b)(a^2 + ab + b^2)$$

7. Factor: $x^3 + 27$

Objective G Review of factoring polynomials in general. (Problem Set exercises 39 – 50 and 59 – 78 are similar.)

8. Factor: $12x^5 - 27x^3$

9. Factor: $5x^4 - 30x^3 + 45x^2$

10. Factor: $3y^2 + 75$

11. Factor: $12a^2 - 13a - 4$

12. Factor: $6x^4 + 6x$

13. Factor: $9x^4y - 18x^3y + 9x^2y^3$

14. Factor: $8ax - 24x + 10a - 30$

15. Factor: $5x^2(x+2) - 8x(x+2) - 4(x+2)$

Matched Problems with Objectives Name ____________________

Quotients

Section 1.5 Date ____________________

Directions Each problem below is similar to the example with the same number in your textbook. After reading through an example in your textbook, or watching one of the videos of that example on MathTV, try the matched problem to check your progress in this section. The shaded text is the learning objective associated with the matched problems that appear below the objective.

Objective A Simplify quotients using the rule for order of operations. (Problem Set exercises 1 – 32 are similar.)

1. Simplify: $\frac{-9-6}{-2-3}$

2. Simplify: $\frac{-9(-2)+3(-3)}{-2(-2)-3}$

3. Simplify: $\frac{3^2+2^2}{3^2-2^3}$

Objective B Use the properties of exponents to simplify expressions containing negative integer exponents. (Problem Set exercises 39 – 72 are similar.)

4. Write 4^{-2} with a positive exponent, then simplify.

5. Write $(-3)^{-3}$ with a positive exponent, then simplify.

6. Write $\left(\frac{4}{5}\right)^{-2}$ with a positive exponent, then simplify.

7. Simplify each of the following. Write answers with positive exponents only:

a. $\frac{2^7}{2^5}$

b. $\frac{a^5}{a^{-5}}$

c. $\frac{x^{11}}{x^{15}}$

d. $\frac{m^{-6}}{m^{-4}}$

Matched Problems with Objectives Name ____________________

Quotients

Section 1.5 Date ____________________

SECTION 1.5

8. Simplify each of the following:

a. $(3x^6y^5)^0$

b. $(3x^6y^5)^1$

9. Simplify: $\dfrac{(x^2)^{-4}(x^3)^2}{(x^{-4})^3}$

10. Simplify: $\dfrac{5a^3b^{-4}}{20a^{-5}b^3}$

11. Simplify: $\dfrac{(2x^{-6}y^2)^3}{(x^2y^{-4})^{-2}}$

12. A solution of hydrochloric acid (HCl) and water contains 36 milliliters of water and 24 milliliters of HCl. Find the ratio of HCl to water and of HCl to the total volume of the solution.

Objective C Find the value of expressions containing quotients. (Problem Set exercises 77 – 80 are similar.)

13. Find the value of each expression when $x = 4$.

a. $\dfrac{x^3+8}{x^2-4}$ b. $\dfrac{x^2-2x+4}{x-2}$ c. $x-2$ d. $x+2$

Matched Problems with Objectives Name ______________________

Unit Analysis and Scientific Notation

Section 1.6 Date ______________________

Directions Each problem below is similar to the example with the same number in your textbook. After reading through an example in your textbook, or watching one of the videos of that example on MathTV, try the matched problem to check your progress in this section. The shaded text is the learning objective associated with the matched problems that appear below the objective.

Objective A Use the appropriate conversion factor to convert units. (Problem Set exercises 1 – 16 are similar.)

1. A rider on a Ferris Wheel would travel approximately 38.6 feet per minute. Convert 38.6 feet per minute to miles per hour.

2. A ski resort advertised that their new chair lift had a speed of 950 feet per second. Show why this speed cannot be correct.

3. If there is one death every 13 seconds in the United States, what is the average number of deaths in one week?

4. At a certain location the pacific plate is moving 420 kilometers every 5 million years. What is the rate of movement in cm/year?

5. A volcanic eruption travels at a speed of 40 meters/second. Convert this speed to miles per hour.

6. A bottle of calcium contains 100 tablets. Each tablet contains 600 milligrams of calcium. What is the total number of grams of calcium in the bottle?

Matched Problems with Objectives Name ____________________

Unit Analysis and Scientific Notation

Section 1.6 Date ____________________

7. The engine in a car has a 2.5-liter displacement. What is the displacement in cubic inches?

Objective B Use scientific notation to write numbers and to simplify expressions. (Problem Set exercises 17 – 46 are similar.)

8. Write 56,700 in scientific notation.

9. Write 3.89×10^4 in expanded form.

10. Multiply: $(5 \times 10^8)(3 \times 10^{-4})$

11. Divide: $\dfrac{8.1 \times 10^{12}}{3 \times 10^7}$

12. Simplify: $\dfrac{(1.2 \times 10^4)(6.4 \times 10^{-2})}{3.2 \times 10^{-6}}$

Matched Problems with Objectives Name ______________________

Linear and Quadratic Equations

Section 2.1 Date ______________________

Directions Each problem below is similar to the example with the same number in your textbook. After reading through an example in your textbook, or watching one of the videos of that example on MathTV, try the matched problem to check your progress in this section. The shaded text is the learning objective associated with the matched problems that appear below the objective.

Objective A Solve a linear equation in one variable. (Problem Set exercises 1– 8, 23, 24, 27– 34, and 39– 44 are similar.)

1. Solve: $4a + 7 = -3a + 28$

2. Solve: $\frac{4}{5}x + \frac{2}{15} = -\frac{3}{10}$

3. Solve: $0.03x + 0.05(1,000 - x) = 43$

4. Solve: $10 - 2(3x - 4) = 48 - 5x$

Objective B Solve equations by factoring. (Problem Set exercises 9 – 18, 25, 26, 35-37, 45, and 46 are similar.)

5. Solve: $x^2 - 8x + 15 = 0$

6. Solve: $90x^2 = 10x$

Matched Problems with Objectives Name ______________________

Linear and Quadratic Equations

Section 2.1 Date ______________________

7. Solve: $(x+5)(x-6)=12$

8. Solve: $6x^3-6x=16x^2$

9. Solve: $x^3-3x^2-4x+12=0$

Objective C Recognize and solve identities and equations with no solution. (Problem Set exercises 47– 52 are similar.)

10. Solve: $4(2x-5)+3=2(4x+1)$

11. Solve: $5(x-2)-2x=3x-10$

Matched Problems with Objectives Name ____________________

Formulas

Section 2.2 Date ____________________

Each problem below is similar to the example with the same number in your textbook. After reading through an example in your textbook, or watching one of the videos of that example on MathTV, try the matched problem to check your progress in this section. The shaded text is the learning objective associated with the matched problems that appear below the objective.

Objective A Solve a formula with numerical replacements for all but one of its variables. (Problem Set exercises 1 – 28 are similar.)

1. Find y when x is 3 in the formula $4x-3y=24$

2. A store selling art supplies finds that they can sell x sketch pads each week at a price of p dollars each, according to the formula $x=800-40p$. What price should they charge for each sketch pad if they want to sell 740 pads each week?

3. A boat is traveling downstream with the current. If the speed of the boat in still water is r and the speed of the current is c, then the formula for the distance traveled by the boat is $d=(r+c)\cdot t$, where t is the length of time. Find c if $d=72$ miles, $r=16$ miles per hour, and $t=4$ hours.

4. If an object is projected into the air with an initial vertical velocity v (in feet/second), its height h (in feet) after t seconds will be given by $h=vt-16t^2$. Find t if $v=80$ feet/second and $h=64$ feet.

5. A manufacturer of small calculators know that the number of calculators it can sell each week is related to the price of the calculators by the equation $x=1200-100p$, where x is the number of calculators and p is the price per calculator. What price should she charge for each calculator if she wants the weekly revenue to be $3,600?

Matched Problems with Objectives Name ____________________

Formulas

Section 2.2 Date ____________________

Objective B Solve formulas for the indicated variable. (Problem Set exercises 29 – 74 are similar.)

6. Given the formula $P = 2x + 2y + 2z$, solve for x.

7. A Ferris wheel has a diameter of 275 feet. One trip around the wheel takes 25 minutes. Find the average speed of a rider on this Ferris wheel. (Use 3.14 as an approximation for π.)

8. Solve for x: $ax - 4 = 2bx + 8$

9. Solve for y: $ax + by = c$

10. Solve for y: $\frac{y+5}{x-3} = 4$

11. Solve the formula $A = P + Prt$ for r.

Matched Problems with Objectives Name ______________________

Applications

Section 2.3 Date ______________________

Directions Each problem below is similar to the example with the same number in your textbook. After reading through an example in your textbook, or watching one of the videos of that example on MathTV, try the matched problem to check your progress in this section. The shaded text is the learning objective associated with the matched problems that appear below the objective.

Objective A Apply the Blueprint for Problem Solving to a variety of application problems. (Problem Set exercises 1 – 52 are similar.)

1. The length of a rectangle is 4 inches less than twice the width. The perimeter is 70 inches. Find the length and width.

2. In July, Ken bought a used Toyota Sienna. The total price, which includes the price of the car plus sales tax, was \$19,305. If the sales tax rate was 7.25%, what was the price of the van?

3. Two complementary angles are such that one is three more than twice the other. Find the measure of the two angles.

4. Suppose a person invests a total of \$10,000 in two accounts. One account earns 5% annually and other earns 3% annually. If the total interest earned from both accounts in a year is \$410, how much is invested in each account?

Matched Problems with Objectives Name ____________________

Applications

Section 2.3 Date ____________________

5. The lengths of the three sides of a right triangle are three consecutive even integers. Find the lengths of the three sides.

6. Two boats leave from an island port at the same time. One travels due north at a speed of ten miles per hour. The other travels due west at a speed of twenty-four miles per hour. How long until the distance between the two boats is 65 miles?

Objective B Use a formula to construct a table of paired data. (Problem Set exercises 53 – 60 are similar.)

7. A piece of string 14 inches long is to be formed into a rectangle. Build a table that gives the length of the rectangle if the width is 1, 2, 3, 4, 5, or 6 inches. Then find the area of each of the rectangles formed.

Matched Problems with Objectives
Linear Equalities in One Variable
Section 2.4

Name ______________________

Date ______________________

Directions Each problem below is similar to the example with the same number in your textbook. After reading through an example in your textbook, or watching one of the videos of that example on MathTV, try the matched problem to check your progress in this section. The shaded text is the learning objective associated with the matched problems that appear below the objective.

Objective A Solve a linear inequality in one variable and then graph the solution set. (Problem Set exercises 1 – 30 are similar.)

1. Solve the following inequality and write the answer using interval notation: $5x+7>3x-9$

Objective B Solve a linear inequality in one variable and then write the solution set using interval notation. (Problem Set exercises 31 – 46 are similar.)

2. Solve the following inequality and write the answer using interval notation: $-6x+5\geq 4x-15$

3. Solve the following inequality and write the answer using interval notation: $3(5x-1)-7x<3(x+4)$

Objective C Solve a compound and continued inequalities and then graph the solution set. (Problem Set exercises 47 – 64 are similar.)

4. Graph: $x\leq 3$ or $x>5$

5. Graph: $x\geq -3$ and $x<2$

Matched Problems with Objectives
Linear Equalities in One Variable
Section 2.4

Name ______________________

Date ______________________

SECTION 2.4

6. Solve and graph: $-4 \le 3x + 2 \le 11$

7. Solve the compound inequality:
$2x - 1 \le -7$ or $2x - 1 \ge 7$

Objective D Solve application problems using inequalities. (Problem Set exercises 67 – 77 are similar.)

8. A company that manufactures ink cartridges for printers finds that they can sell x cartridges each week at a price of p dollars each, according to the formula $x = 900 - 60p$. What price should they charge for each cartridge if they want to sell at least 180 cartridges a week?

9. The formula $F = \frac{9}{5}C + 32$ gives the relationship between the Celsius and Fahrenheit temperature scales. If the temperature range on a certain day is $59°$ F to $77°$ F, what is the temperature range in degrees Celsius?

Matched Problems with Objectives Name ____________________

Equations with Absolute Values

Section 2.5 Date ____________________

Directions Each problem below is similar to the example with the same number in your textbook. After reading through an example in your textbook, or watching one of the videos of that example on MathTV, try the matched problem to check your progress in this section. The shaded text is the learning objective associated with the matched problems that appear below the objective.

Objective A Solve equations with absolute value symbols. (Problem Set exercises 1 – 70 are similar.)

1. Solve for x: $|x| = 6$

2. Solve for x: $|2x+1| = 11$

3. Solve for x: $\left|\frac{2}{5}x+7\right| - 1 = 2$

4. Solve for x: $|3x-1| = -5$

5. Solve for x: $|2x+1| = |x+2|$

6. Solve for x: $|x+3| = |x-6|$

Matched Problems with Objectives Name ______________________

Inequalities Involving Absolute Value

Section 2.6 Date ______________________

Directions Each problem below is similar to the example with the same number in your textbook. After reading through an example in your textbook, or watching one of the videos of that example on MathTV, try the matched problem to check your progress in this section. The shaded text is the learning objective associated with the matched problems that appear below the objective.

Objective A Solve inequalities with absolute value and graph the solution set. (Problem Set exercises 1 – 70 are similar.

1. Solve and graph: $|3x-1|<8$

2. Solve and graph: $|2x+5|\le 13$

3. Solve and graph: $|x-7|>5$

4. Solve and graph: $|5x-6|\ge 4$

5. Solve and graph: $|4x+3|-2<5$

6. Solve and graph: $|6-3x|>15$

7. Solve: $|3x+7|<-2$

8. Solve: $|7x+6|>-8$

Matched Problems with Objectives

Paired Data and the Rectangular Coordinate System

Section 3.1

Name ______________________

Date ______________________

SECTION 3.1

Directions Each problem below is similar to the example with the same number in your textbook. After reading through an example in your textbook, or watching one of the videos of that example on MathTV, try the matched problem to check your progress in this section. The shaded text is the learning objective associated with the matched problems that appear below the objective.

Objective A Graph ordered pairs on a rectangular coordinate system. (Problem Set exercises 1 – 4 are similar.)

1. Plot (graph) the ordered pairs
$(1,3),(-1,3),(-1,-3)$, and $(1,-3)$.

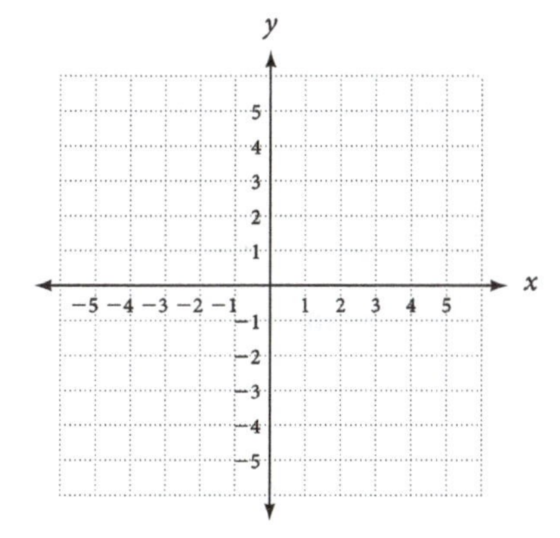

2. Plot (graph) the ordered pairs

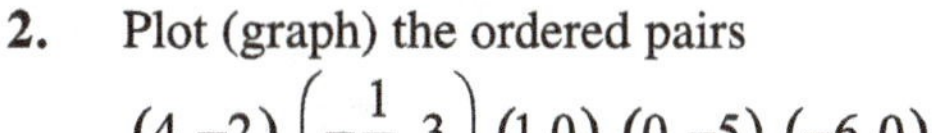
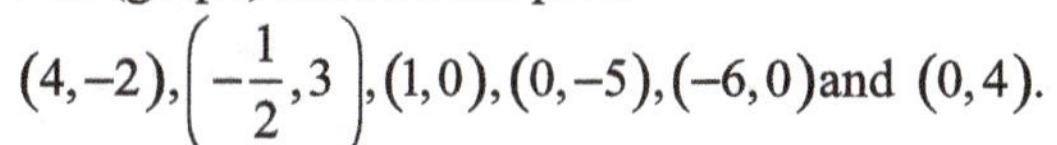
$(4,-2),\left(-\frac{1}{2},3\right),(1,0),(0,-5),(-6,0)$ and $(0,4)$.

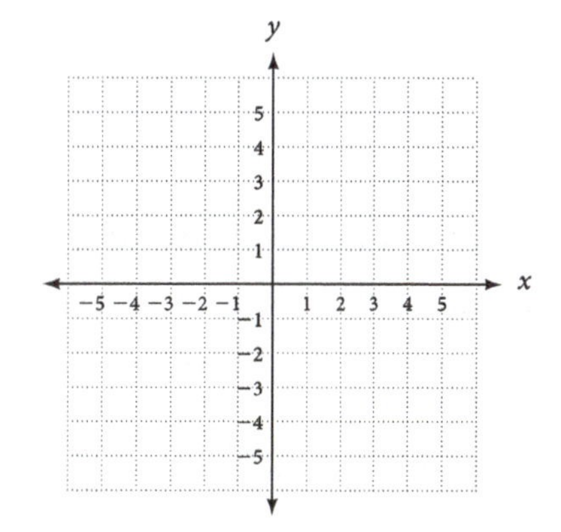

Objective B Graph linear equations by making a table. (Problem Set exercises 5 – 8 are similar.)

3. Graph $y=\frac{1}{2}x$

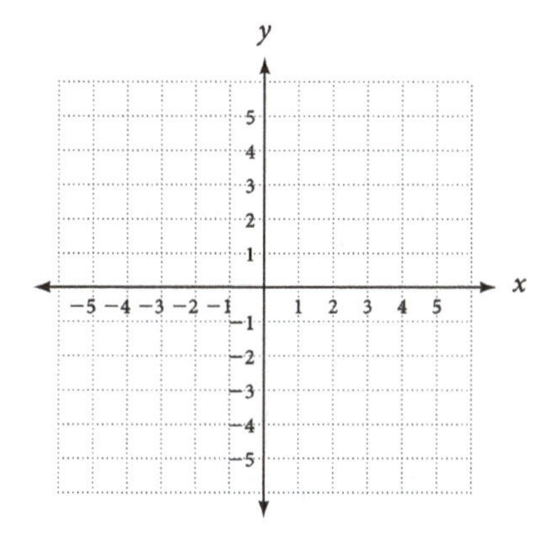

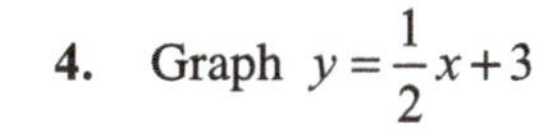
4. Graph $y=\frac{1}{2}x+3$

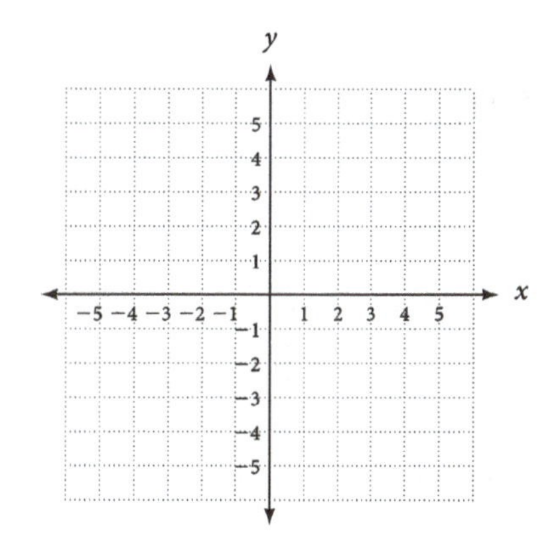

Matched Problems with Objectives

Paired Data and the Rectangular Coordinate System

Section 3.1

Name ______________________

Date ______________________

5. Graph $y = \frac{1}{2}x - 4$

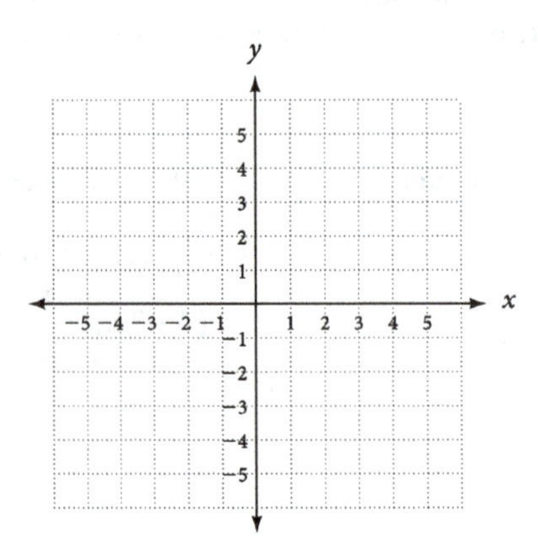

Objective C Graph equations using vertical translations. (Problem Set exercises 9 – 16 are similar.)

6. Find an equation for the line.

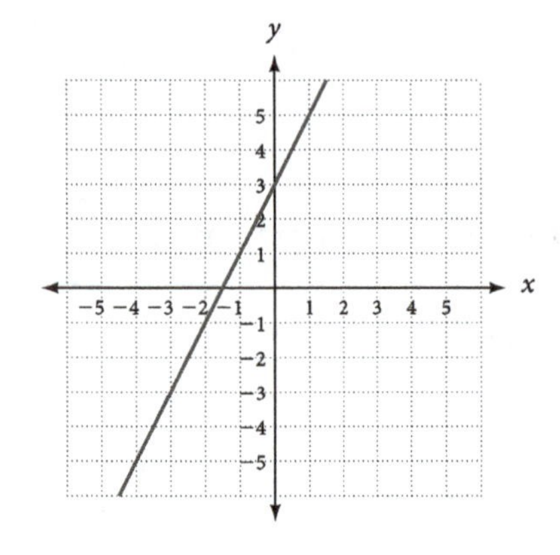

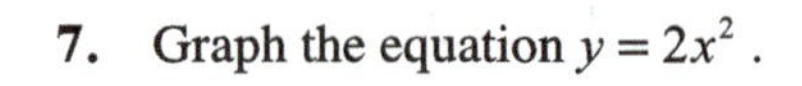

7. Graph the equation $y = 2x^2$.

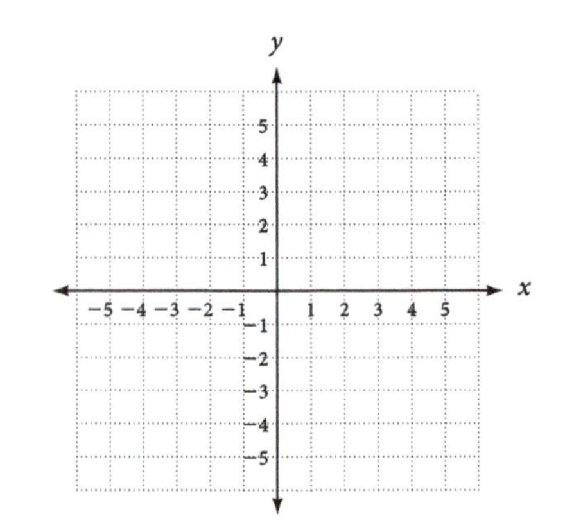

8. Graph the equation $y = 2x^2 - 4$.

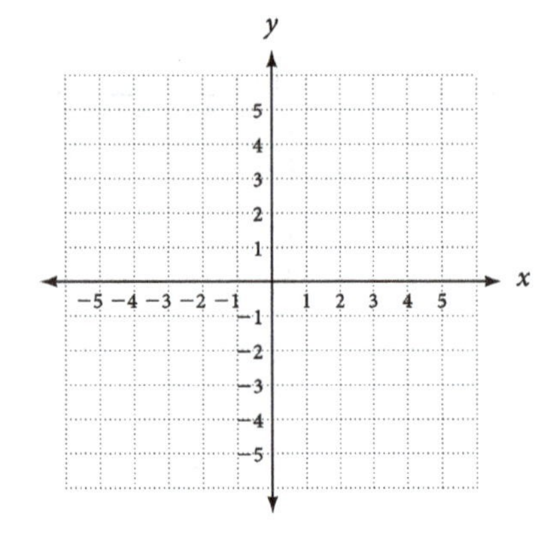

9. Graph the equation $y = |x|$ and $y = |x| + 2$.

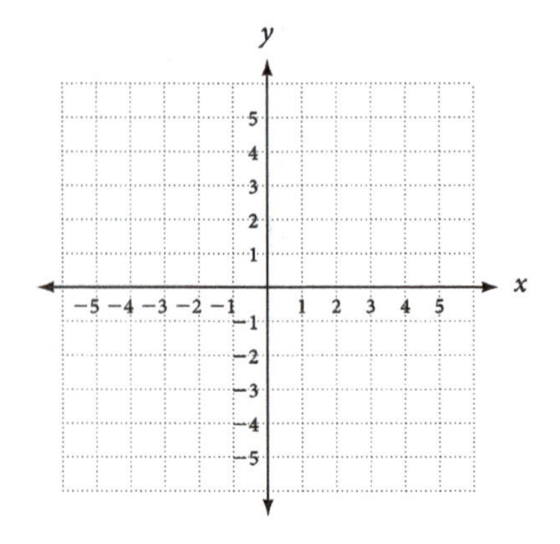

Matched Problems with Objectives Name ______________________

Paired Data and the Rectangular Coordinate System

Section 3.1 Date ______________________

Objective D Graph equations using intercepts. (Problem Set exercises 17, 18, and 23 – 36 are similar.)

10. Find the *x*- and *y*- intercepts for $2x-3y=6$, then graph the solution set.

11. Find the intercepts for $y=x^2-1$.

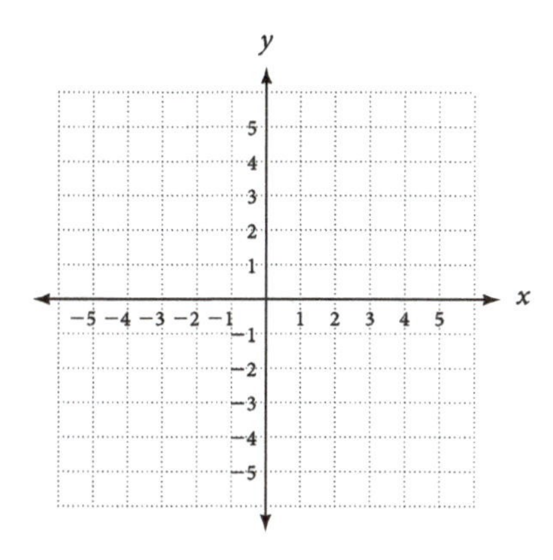

Objective E Graph horizontal and vertical lines. (Problem Set exercises 19 – 22 are similar.)

12. Graph each of the following lines: $x=-1$ and $y=4$

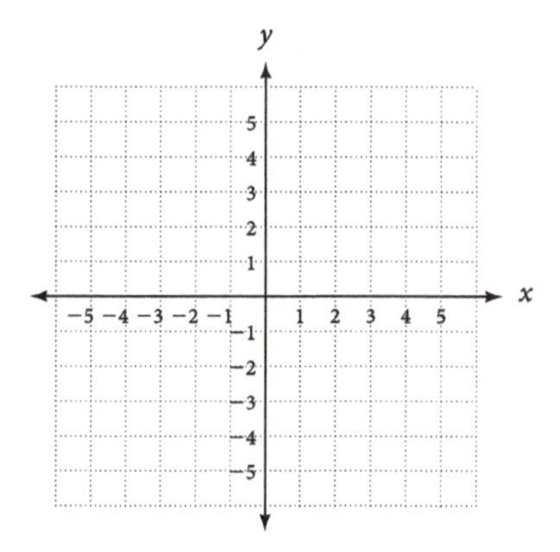

Matched Problems with Objectives Name ______________________

The Slope of a Line

Section 3.2 Date ______________________

Directions Each problem below is similar to the example with the same number in your textbook. After reading through an example in your textbook, or watching one of the videos of that example on MathTV, try the matched problem to check your progress in this section. The shaded text is the learning objective associated with the matched problems that appear below the objective.

Objective A Find the slope of a line from its graph. (Problem Set exercises 1 – 6 are similar.)

1. Graph the line $y = 3x - 5$ and then find its slope.

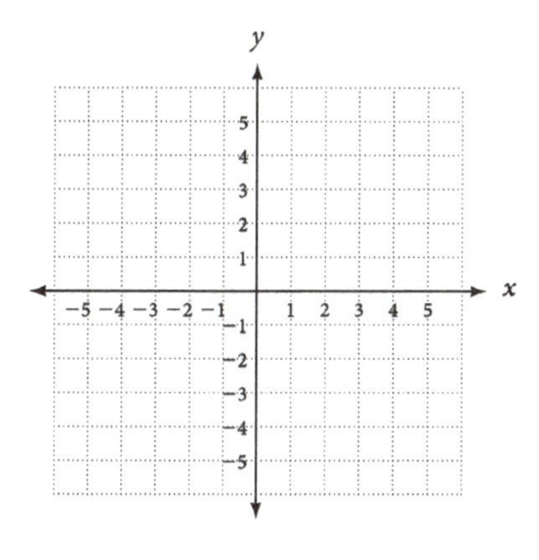

Objective B Find the slope of a line given two points on the line. (Problem Set exercises 7 – 34 are similar.)

2. Find the slope of the line through $(3,4)$ and $(1,-2)$.

3. Find the slope of the line through $(2,-3)$ and $(-1,-3)$.

Matched Problems with Objectives Name ____________________

The Slope of a Line

Section 3.2 Date ____________________

SECTION 3.2

4. On the chart in the textbook on Page 167, find the slope of the line connecting the point $(1985, 1.25)$ with the last point $(2005, 2.93)$. Explain in words what your result represents.

Objective C Use a graph to find the average speed of an object. (Problem Set exercises 45 – 48 are similar.)

5. An object is traveling at a constant speed. The distance and time data are shown on the graph below. Use the graph to find the speed of the object.

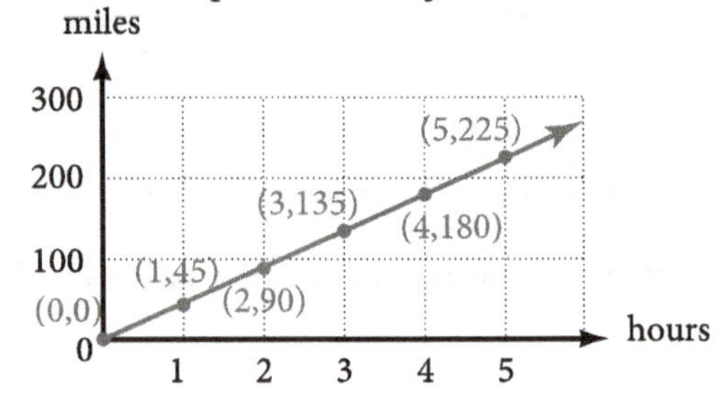

Objective D Find the slope of a line parallel or perpendicular to a given line. (Problem Set exercises 35 – 42 are similar.)

6. Find the slope of any line perpendicular to the line through the points $(2, -3)$ and $(6, 5)$.

Matched Problems with Objectives Name ____________________

The Equation of a Line

Section 3.3 Date ____________________

Directions Each problem below is similar to the example with the same number in your textbook. After reading through an example in your textbook, or watching one of the videos of that example on MathTV, try the matched problem to check your progress in this section. The shaded text is the learning objective associated with the matched problems that appear below the objective.

Objective A Find the equation of a line given its slope and y-intercept. (Problem Set exercises 1 – 6 are similar.)

1. Find the equation of the line with slope $\frac{2}{3}$ and y-intercept 1. Then graph the line.

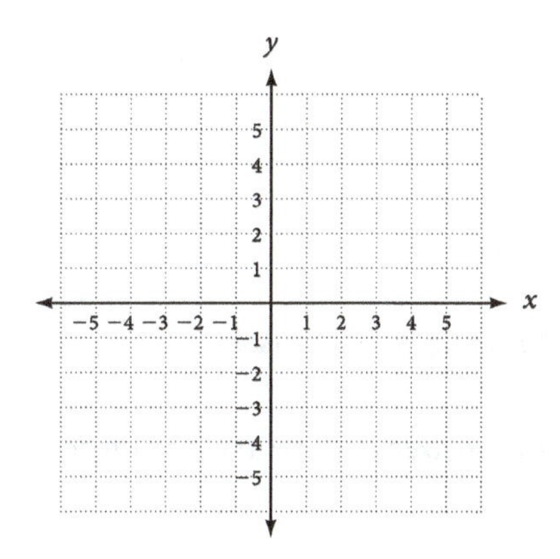

Objective B Find the slope and y-intercept from the equation of a line. (Problem Set exercises 13 –22 are similar.)

2. Give the slope and y-intercept for the line $4x-5y=7$.

Matched Problems with Objectives Name ______________________

The Equation of a Line

Section 3.3 Date ______________________

3. Give the slope and y-intercept for the line $-3x+2y=-6$. Then use the slope and y-intercept to graph the line.

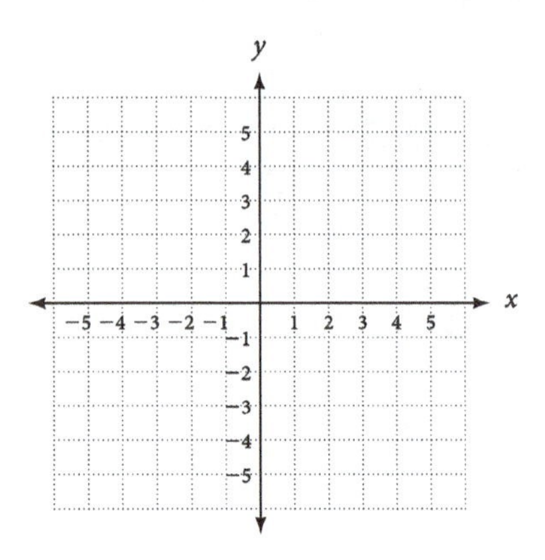

Objective C Find the equation of a line given its slope and a point on the line. (Problem Set exercises 23 – 32 are similar.)

4. Find the equation of the line with slope 3 that contains the point $(-1,2)$. Write the answer in slope-intercept form.

Matched Problems with Objectives Name ______________________

The Equation of a Line

Section 3.3 Date ______________________

Objective D Find the equation of a line given two points on the line. (Problem Set exercises 33 – 42 are similar.)

5. Find the equation of the line that contains the points $(2,5)$ and $(6,-3)$. Write the answer in slope-intercept form.

Objective E Find the equation of a line parallel or perpendicular to a given line and through a given point. Write answer in standard form. (Problem Set exercises 7 – 12 and 49 – 53 are similar.)

6. Find the equation of the line through (3, 2) that is perpendicular to the graph of $3x - y = 2$. Write your answer in standard form.

Matched Problems with Objectives

Name ______________________

Linear Inequalities in Two Variables

Section 3.4

Date ______________________

Directions Each problem below is similar to the example with the same number in your textbook. After reading through an example in your textbook, or watching one of the videos of that example on MathTV, try the matched problem to check your progress in this section. The shaded text is the learning objective associated with the matched problems that appear below the objective.

Objective A Graph linear inequalities in two variables. (Problem Set exercises 1 – 40 are similar.)

1. Graph the solution set for $x - y \geq 3$.

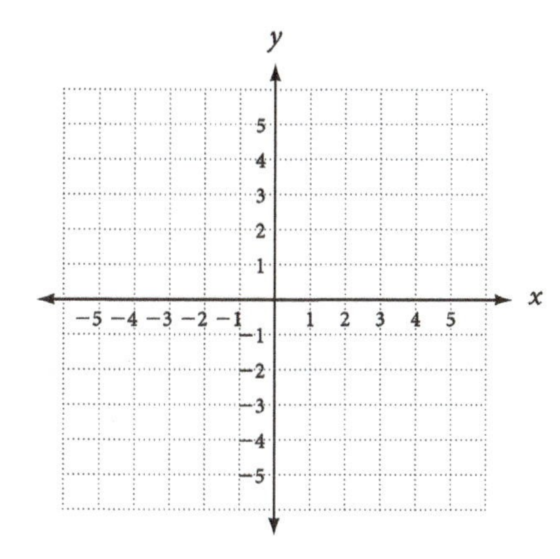

2. Graph the solution set for $y < \frac{1}{2}x + 3$.

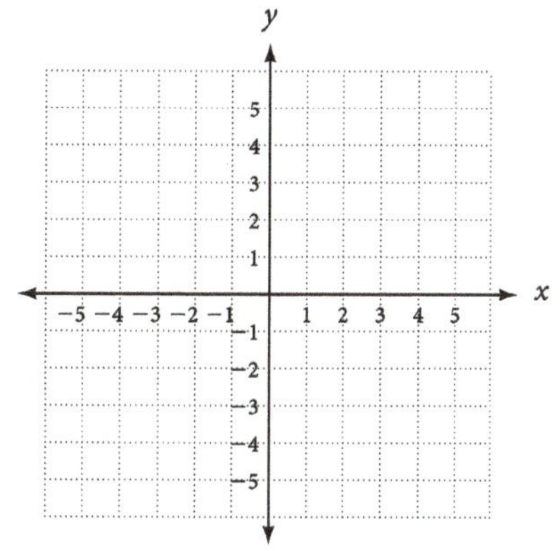

3. Graph the solution set for $y > -2$.

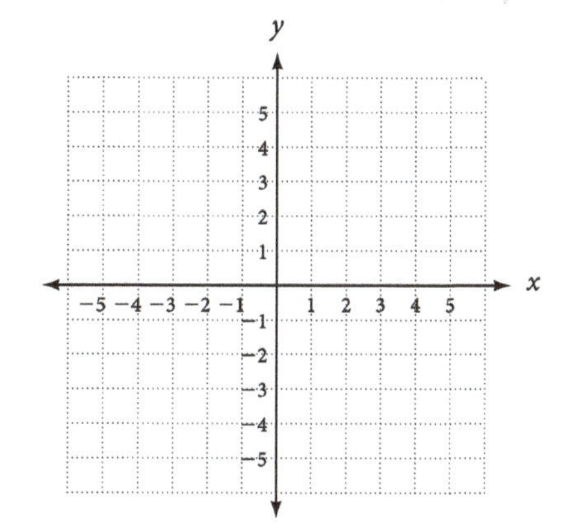

Matched Problems with Objectives Name ____________________

Introduction to Functions

Section 3.5 Date ____________________

Directions Each problem below is similar to the example with the same number in your textbook. After reading through an example in your textbook, or watching one of the videos of that example on MathTV, try the matched problem to check your progress in this section. The shaded text is the learning objective associated with the matched problems that appear below the objective.

Objective A Construct a table or a graph from a function rule. (Problem Set exercises 33 – 34 are similar.)

1. Construct a table and graph for $y = 8x, 0 \le x \le 40$.

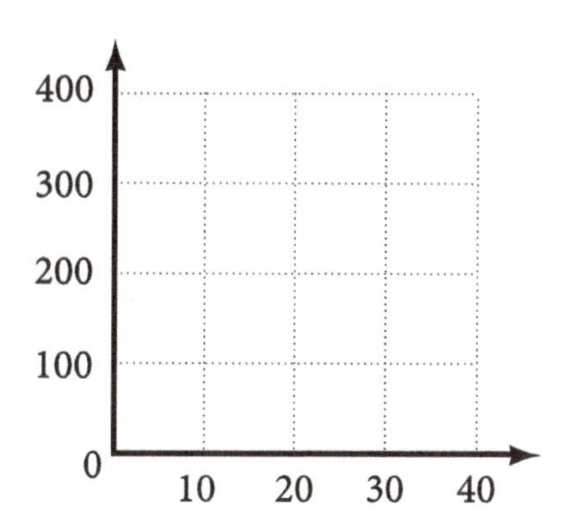

Objective B Identify the domain and range of a function or relation. (Problem Set exercises 1 – 8 and 19 – 22 are similar.)

2. State the domain and range for $y = 8x, 0 \le x \le 40$.

3. Kendra is tossing a softball into the air so that the distance h the ball is above her hand t seconds after she begins the toss is given by $h = 48t - 16t^2, 0 \le t \le 3$. Construct a table and line graph for this function.

Time (sec) t	Function Rule $h = 48t - 16t^2$	Distance h
0		
1		
1.5		
2		
3		

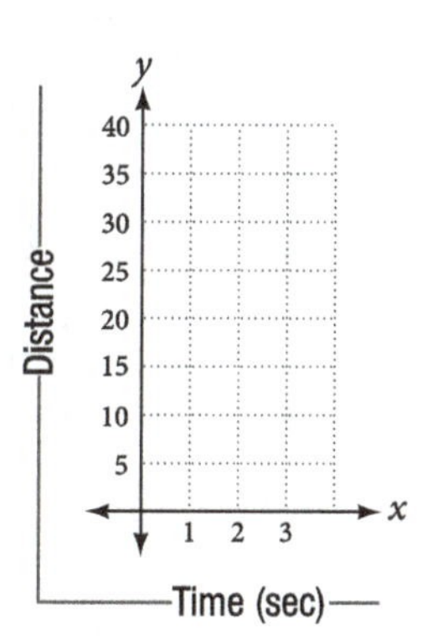

Matched Problems with Objectives

Introduction to Functions
Section 3.5

Name ____________________

Date ____________________

Objective C Determine whether a relation is also a function. (Problem Set exercises 9 – 18 and 23 – 32 are similar.)

4. Graph the following set of ordered pairs and then determine whether this relation is a function:

$$\{(1,5),(2,4),(4,8),(2,9),(5,10),(6,12)\}$$

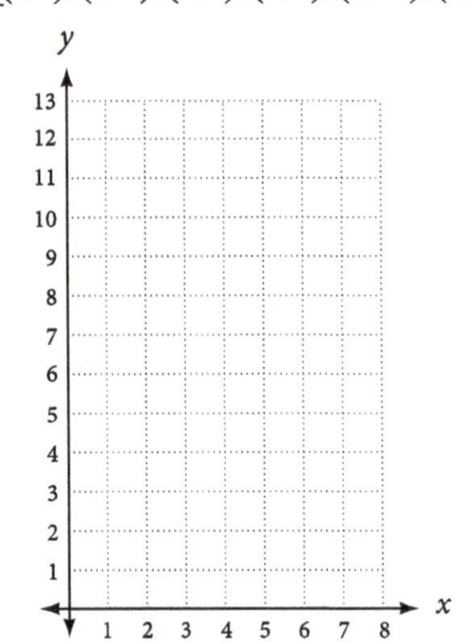

5. Use the equation $x = y^2 - 4$ to fill in the following table. Then use the table to sketch the graph. Is this graph a function?

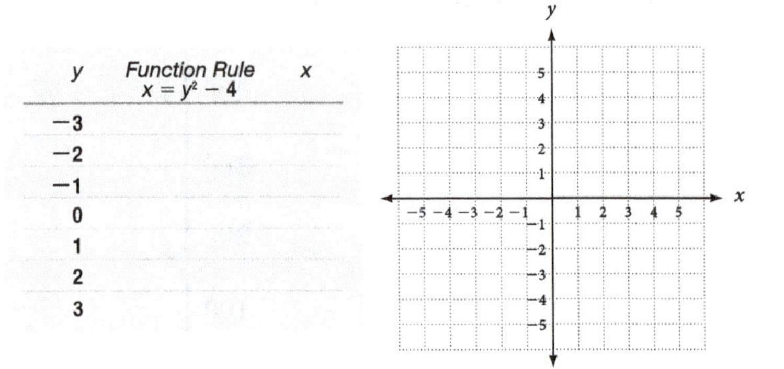

y	Function Rule $x = y^2 - 4$	x
−3		
−2		
−1		
0		
1		
2		
3		

6. Graph $y = |x| - 3$. Is this the graph of a function?

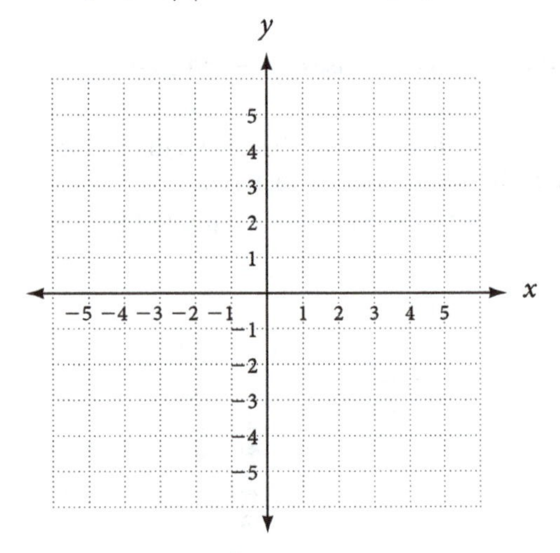

Matched Problems with Objectives Name ____________________

Function Notation

Section 3.6 Date ____________________

Directions Each problem below is similar to the example with the same number in your textbook. After reading through an example in your textbook, or watching one of the videos of that example on MathTV, try the matched problem to check your progress in this section. The shaded text is the learning objective associated with the matched problems that appear below the objective.

Objective A Use function notation to find the value of a function for a given value of the variable. (Problem Set exercises 1 – 46 are similar.)

1. If $f(x)=8x$, find each of the following:

a. $f(0)$

b. $f(5)$

c. $f(10.5)$

2. When Lorena runs a mile in t minutes, then her average speed in feet per seconds is given by

$s(t)=\frac{88}{t},\ t>0$.

a. Find $s(8)$ and explain what it means.

b. Find $s(11)$ and explain what it means.

3. A medication has a half-life of 5 days. If the concentration of the medication in a patient's system is 80 ng/mL, and the patient stops taking it, then t days later the concentration will be $C(t)=80\left(\frac{1}{2}\right)^{t/5}$. Find each of the following and explain what they mean.

a. $C(5)$

b. $C(10)$

4. The following formulas give the circumference and area of a circle with a radius of r. Use the formulas to find the circumference and area of a circular plate if the radius is 5 inches.

$$C(r)=2\pi r$$

$$A(r)=\pi r^2$$

Matched Problems with Objectives Name ______________________

Function Notation

Section 3.6 Date ______________________

5. If $f(x) = 4x^2 - 3$, find the following:

a. $f(0)$

b. $f(3)$

c. $f(-2)$

6. If $f(x) = 2x + 1$ and $g(x) = x^2 - 3$, find the following:

a. $f(5)$

b. $g(5)$

c. $f(-2)$

d. $g(-2)$

e. $f(a)$

f. $g(a)$

7. If $f = \{(-4,1),(2,-3),(7,9)\}$, find the following:

a. $f(-4)$

b. $f(2)$

c. $f(7)$

8. If $f(x) = 3x^2$ and $g(x) = 4x + 1$, find the following:

a. $f[g(2)]$

b. $g[f(2)]$

Matched Problems with Objectives

Variation

Section 3.7

Name ______________________

Date ______________________

Directions Each problem below is similar to the example with the same number in your textbook. After reading through an example in your textbook, or watching one of the videos of that example on MathTV, try the matched problem to check your progress in this section. The shaded text is the learning objective associated with the matched problems that appear below the objective.

Objective A Set up and solve problems with direct, inverse, or joint variation. (Problem Set exercises 1 – 33 are similar.)

1. y varies directly with x. If y is 24 when x is 8, find y when x is 2.

2. A skydiver jumps from a plane. Like any object that falls toward Earth, the distance the skydiver falls is directly proportional to the square of the time he has been falling until he reaches his terminal velocity. If the skydiver falls 64 feet in the first 2 seconds of the jump, then how far will he have fallen after 2.5 seconds?

3. The volume of a gas is inversely proportional to the pressure of the gas on its container. If a pressure of 48 pounds per square inch corresponds to a volume of 50 cubic feet, what pressure is needed to produce a volume of 150 cubic feet?

4. y varies jointly with x and the square of z. If y is 81 when x is 2 and z is 9, find y when x is 4 and z is 4.

5. In electricity, the resistance of a cable is directly proportional to its length and inversely proportional to the square of the diameter. If a 100-foot cable 0.5 inch in diameter has a resistance of 0.2 ohm, what will be the resistance of a 300-foot cable made of the same material that is 0.25 inch in diameter?

Matched Problems with Objectives

Name ______________________

Algebra and Composition with Functions

Section 3.8

Date ______________________

Directions Each problem below is similar to the example with the same number in your textbook. After reading through an example in your textbook, or watching one of the videos of that example on MathTV, try the matched problem to check your progress in this section. The shaded text is the learning objective associated with the matched problems that appear below the objective.

Objective A Find the sum, difference, product, and quotient of functions. (Problem Set exercises 1 – 38 are similar.)

1. If $f(x)=x^2-4$ and $g(x)=x+2$, find formulas for the following:

a. $f+g$

b. $f-g$

c. fg

d. $\frac{f}{g}$

2. If $f(x)=3x+2$, $g(x)=3x^2-10x-8$, and $h(x)=x-4$, find formulas for the following:

a. $f+g$

b. fh

c. fg

d. $\frac{g}{f}$

3. If $f(x)=4x-3$, $g(x)=4x^2-7x+3$, and $h(x)=x-1$, find the following:

a. $(f+g)(3)$

b. $(fh)(-1)$

c. $(fg)(0)$

d. $\left(\frac{g}{f}\right)(10)$

Objective B Find the composition of functions. (Problem Set exercises 39 – 44 are similar.)

4. If $f(x)=x-4$ and $g(x)=x^2+3x$, find the following:

a. $(f\circ g)(x)$

b. $(g\circ f)(x)$

Matched Problems with Objectives

Name ____________________

Systems of Linear Equations in Two Variables

Section 4.1

Date ____________________

Directions Each problem below is similar to the example with the same number in your textbook. After reading through an example in your textbook, or watching one of the videos of that example on MathTV, try the matched problem to check your progress in this section. The shaded text is the learning objective associated with the matched problems that appear below the objective.

Objective A Solve systems of linear equations in two variables by the addition method. (Problem Set exercises 9 –24 and 35 – 40 are similar.)

1. Solve the system using the addition method:

$$2x - y = 7$$
$$3x + 4y = -6$$

2. Solve the system using the addition method:

$$3x - 2y = -8$$
$$-2x + 3y = 7$$

SECTION 4.1

3. Solve the system using the addition method:

$$3x - 5y = 2$$
$$2x + 4y = 1$$

4. Solve the system using the addition method:

$$2x + 7y = 3$$
$$4x + 14y = 1$$

Matched Problems with Objectives

Name ____________________

Systems of Linear Equations in Two Variables

Section 4.1

Date ____________________

SECTION 4.1

5. Solve the system using the addition method:

$$2x + 7y = 3$$
$$4x + 14y = 6$$

6. Solve the system using the addition method:

$$\frac{1}{3}x + \frac{1}{2}y = 4$$
$$\frac{2}{3}x - \frac{1}{4}y = 3$$

Objective B Solve systems of linear equations in two variables by the substitution method. (Problem Set exercises 25 – 34 are similar.)

7. Solve the system using the substitution method:

$$4x - 2y = -2$$
$$y = x + 3$$

8. Solve the system using the substitution method:

$$5x - 3y = -4$$
$$x + 2y = 7$$

Matched Problems with Objectives

Name ______________________

Systems of Linear Equations in Two Variables

Section 4.1

Date ______________________

Objective C Solve systems of linear equations in two variables by graphing. (Problem Set exercises 1 – 8 are similar.)

9. Solve the system by graphing:

$$2x - y = 4$$
$$x + y = 2$$

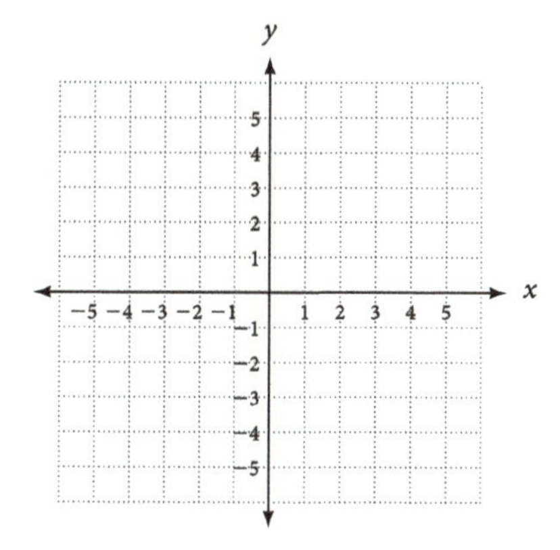

Matched Problems with Objectives Name ____________________

Systems of Linear Equations in Three Variables

Section 4.2 Date ____________________

Directions Each problem below is similar to the example with the same number in your textbook. After reading through an example in your textbook, or watching one of the videos of that example on MathTV, try the matched problem to check your progress in this section. The shaded text is the learning objective associated with the matched problems that appear below the objective.

Objective A Solve systems of linear equations in three variables. (Problem Set exercises 1 –30 are similar.)

1. Solve the system:

$$\begin{aligned} x+2y+z&=2\\ x+y-z&=6\\ x-y+2z&=-7 \end{aligned}$$

2. Solve the system:

$$\begin{aligned} 3x-2y+z&=2\\ 3x+y+3z&=7\\ x+4y-z&=4 \end{aligned}$$

3. Solve the system:

$$\begin{aligned} 3x+5y-2z&=1\\ 6x+10y-4z&=2\\ x-8y+z&=4 \end{aligned}$$

4. Solve the system:

$$\begin{aligned} 3x-y+2z&=4\\ 6x-2y+4z&=2\\ 5x-3y+7z&=5 \end{aligned}$$

5. Solve the system:

$$\begin{aligned} x+2y&=0\\ 3y+z&=-3\\ 2x-z&=5 \end{aligned}$$

Matched Problems with Objectives Name ______________________

Applications

Section 4.3 Date ______________________

Directions Each problem below is similar to the example with the same number in your textbook. After reading through an example in your textbook, or watching one of the videos of that example on MathTV, try the matched problem to check your progress in this section. The shaded text is the learning objective associated with the matched problems that appear below the objective.

Objective A Solve application problems whose solutions are found through systems of linear equations. (Problem Set exercises 1 – 34 are similar.)

1. A number is 1 less than twice another. Their sum is 14. Find the two numbers.

2. There were 750 tickets sold for a basketball game for a total of $1,090. If adult tickets cost $2.00 and children's tickets cost $1.00, how many of each kind were sold?

3. A person invests $12,000 in two accounts. One account earns 6% annually and the other earns 7%. If the total interest in a year is $790, how much was invested in each account?

4. How much 30% alcohol solution and 70% alcohol solution must be mixed to get 16 gallons of 60% solution?

Matched Problems with Objectives Name ______________________

Applications

Section 4.3 Date ______________________

5. A boat can travel 20 miles downstream in 2 hours. The same boat can travel 18 miles upstream in 3 hours. What is the speed of the boat in still water, and what is the speed of the current?

6. A collection of nickels, dimes, and quarters consists of 15 coins with a total value of $1.10. If the number of nickels is one less than 4 times the number of dimes, how many of each coin is contained in the collection?

7. Suppose the students mentioned in Example 7 in the textbook take two additional temperature measurements while they are heating the water and find that when $F = 95°$, $C = 35°$, and when $F = 167°$, $C = 75°$. Assume that the relationship between the two temperature scales is linear, then derive the formula that gives C in terms of F using these new data points.

Matched Problems with Objectives Name ____________________

Systems of Linear Inequalities

Section 4.4 Date ____________________

Directions Each problem below is similar to the example with the same number in your textbook. After reading through an example in your textbook, or watching one of the videos of that example on MathTV, try the matched problem to check your progress in this section. The shaded text is the learning objective associated with the matched problems that appear below the objective.

Objective A Graph the solution to a system of linear inequalities in two variables. (Problem Set exercises 1 – 24 are similar.)

1. Graph the solution set to the system:

$$y \le \frac{1}{3}x + 1$$

$$y > \frac{1}{3}x - 3$$

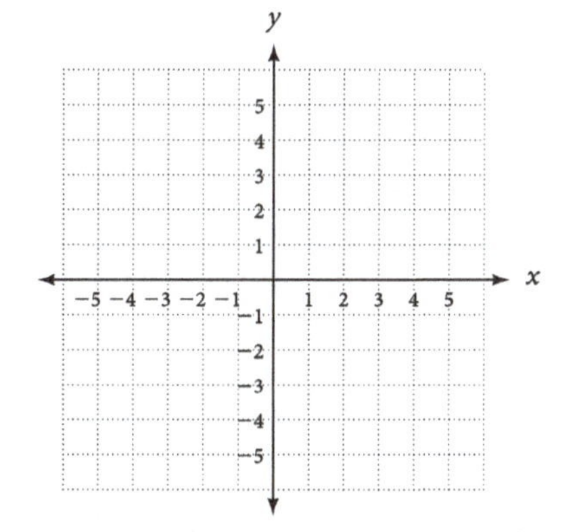

2. Graph the solution set to the system:

$$x + y < 3$$

$$x \ge 0$$

$$y \ge 0$$

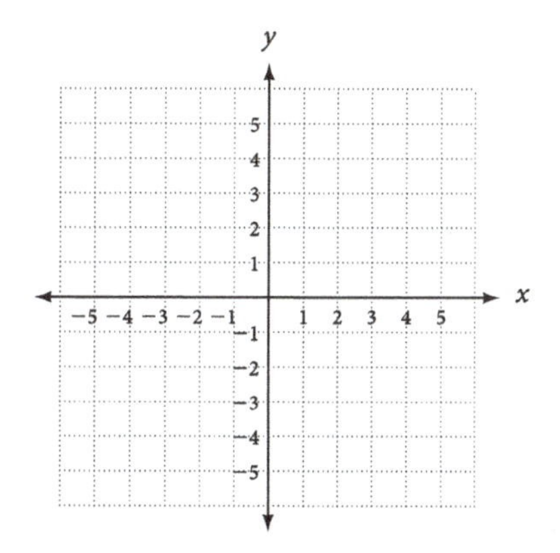

Matched Problems with Objectives Name ____________________

Systems of Linear Inequalities

Section 4.4 Date ____________________

SECTION 4.4

3. Graph the solution set to the system:

$x > -3$

$y < 4$

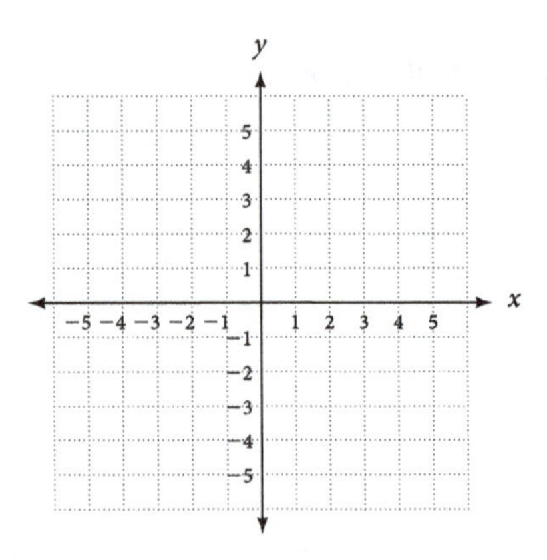

4. Graph the solution set to the system:

$x - y < 5$

$x + y < 5$

$x > 1$

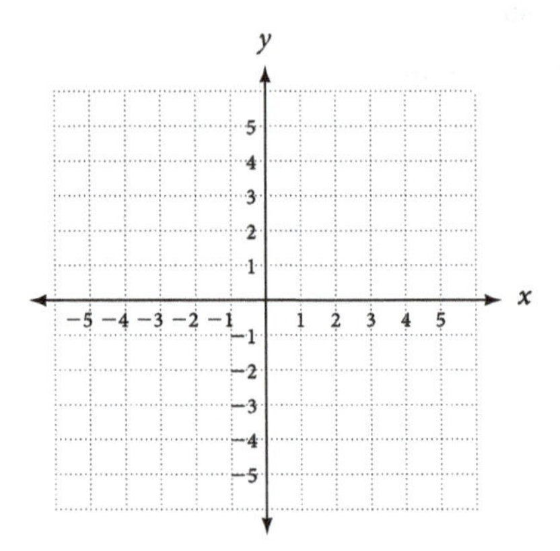

5. How would the graph in Figure 16 in the textbook change if they had reserved only 300 tickets at the $15 rate?

Matched Problems with Objectives

Name ______________________

Basic Properties and Reducing to Lowest Terms

Section 5.1

Date ______________________

Directions Each problem below is similar to the example with the same number in your textbook. After reading through an example in your textbook, or watching one of the videos of that example on MathTV, try the matched problem to check your progress in this section. The shaded text is the learning objective associated with the matched problems that appear below the objective.

Objective A Reduce rational expressions to lowest terms. (Problem Set exercises 11 – 53 are similar.)

1. Reduce to lowest terms: $\dfrac{x^2-9}{x+3}$

2. Reduce to lowest terms: $\dfrac{y^2-y-6}{y^2-4}$

3. Reduce to lowest terms: $\dfrac{3a^3+3}{6a^2-6a+6}$

4. Reduce to lowest terms: $\dfrac{x^2+4x+ax+4a}{x^2+ax+4x+4a}$

5. Reduce to lowest terms: $\dfrac{a-7}{7-a}$

6. Reduce to lowest terms: $\dfrac{7-x}{x^2-49}$

Matched Problems with Objectives

Basic Properties and Reducing to Lowest Terms

Section 5.1

Name ______________________

Date ______________________

Objective B Find function values for rational functions. (Problem Set exercises 1 – 4 and 65 – 66 are similar.)

7. If $f(x)=\dfrac{x+6}{x-3}$, find

a. $f(0)$

b. $f(-6)$

c. $f(6)$

d. $f(3)$

Objective C Work with ratios. (Problem Set exercises 5 – 10 and 55 – 64 are similar.)

8. Find the domain for each function.

a. $f(x)=\dfrac{x-5}{x-1}$

b. $g(x)=\dfrac{x^2+4}{x+2}$

c. $h(x)=\dfrac{x}{x^2-1}$

9. If $f(x)=2x-7$, find $\dfrac{f(x)-f(a)}{x-a}$ and simplify.

10. If $f(x)=x^2-16$, find $\dfrac{f(x)-f(a)}{x-a}$ and simplify.

11. If $f(x)=3x-9$, find $\dfrac{f(x+h)-f(x)}{h}$ and simplify.

Matched Problems with Objectives Name ______________________

Multiplication and Division of Rational Expressions

Section 5.2 Date ______________________

Directions Each problem below is similar to the example with the same number in your textbook. After reading through an example in your textbook, or watching one of the videos of that example on MathTV, try the matched problem to check your progress in this section. The shaded text is the learning objective associated with the matched problems that appear below the objective.

Objective A Multiply and divide rational expressions. (Problem Set exercises 1 – 70 are similar.)

1. Multiply: $\frac{3}{4}\cdot\frac{12}{27}$

2. Multiply: $\frac{6x^4}{4y^9}\cdot\frac{12y^{13}}{3x^2}$

3. Multiply: $\frac{x+5}{x^2-25}\cdot\frac{x-5}{x^2-10x+25}$

4. Multiply: $\frac{3y^2-3y}{3y-12}\cdot\frac{y^2-2y-8}{y^2+3y+2}$

5. Divide: $\frac{5}{9}\div\frac{10}{27}$

6. Divide: $\frac{9x^4}{4y^3}\div\frac{3x^2}{8y^5}$

MathTV.com

Matched Problems with Objectives

Multiplication and Division of Rational Expressions

Section 5.2

Name ______________________

Date ______________________

7. Divide: $\dfrac{xy^2 - y^3}{x^2 - y^2} \div \dfrac{x^3 + y^3}{x^2 + 2xy + y^2}$

8. Perform the indicated operations:

$$\frac{a^2 + 3a - 4}{a - 4} \cdot \frac{a + 3}{a^2 - 4a + 3} \div \frac{a + 1}{a^2 - 2a - 3}$$

SECTION 5.2

9. Multiply: $\dfrac{xa + xb - ya - yb}{xa + 2x + ya + 2y} \cdot \dfrac{xa + 2x + ya + 2y}{xa + xb + ya + yb}$

10. Multiply: $\left(5x^2 - 45\right) \cdot \dfrac{3}{5x - 15}$

Matched Problems with Objectives

Addition and Subtraction of Rational Expressions

Section 5.3

Name ______________________

Date ______________________

Directions Each problem below is similar to the example with the same number in your textbook. After reading through an example in your textbook, or watching one of the videos of that example on MathTV, try the matched problem to check your progress in this section. The shaded text is the learning objective associated with the matched problems that appear below the objective.

Objective A Add and subtract rational expressions with the same denominator. (Problem Set exercises 11 – 22 are similar.)

1. Add: $\frac{3}{8}+\frac{1}{8}$

2. Add: $\frac{x}{x^2-9}+\frac{3}{x^2-9}$

3. Subtract: $\frac{2x-5}{x-2}-\frac{x-3}{x-2}$

Objective B Add and subtract rational expressions with different denominators. (Problem Set exercises 1 – 10 and 25 – 68 are similar.)

4. Add: $\frac{3}{10}+\frac{11}{42}$

5. Add: $\frac{-3}{x^2-2x-8}+\frac{4}{x^2-16}$

Matched Problems with Objectives Name ______________________

Addition and Subtraction of Rational Expressions

Section 5.3 Date ______________________

SECTION 5.3

6. Add: $\dfrac{x-4}{2x-6}+\dfrac{3}{x^2-9}$

7. Subtract: $\dfrac{2x-4}{x^2+5x+4}-\dfrac{x-4}{x^2+6x+8}$

8. Add: $\dfrac{x^2}{x-4}+\dfrac{x+12}{4-x}$

9. Add: $2+\dfrac{25}{5x-1}$

10. One number is three times another. Write an expression for the sum of the reciprocals of the two numbers. Then simplify that expression.

Matched Problems with Objectives

Complex Fractions

Section 5.4

Name ____________________

Date ____________________

Directions Each problem below is similar to the example with the same number in your textbook. After reading through an example in your textbook, or watching one of the videos of that example on MathTV, try the matched problem to check your progress in this section. The shaded text is the learning objective associated with the matched problems that appear below the objective.

Objective A Simplify complex fractions. (Problem Set exercises 1 – 50 are similar.)

1. Simplify:

$$\frac{\frac{2}{3}}{\frac{5}{6}}$$

2. Simplify:

$$\frac{\frac{1}{x}-\frac{1}{3}}{\frac{1}{x}+\frac{1}{3}}$$

3. Simplify:

$$\frac{\frac{x+5}{x^2-16}}{\frac{x^2-25}{x-4}}$$

4. Simplify:

$$\frac{1-\frac{9}{x^2}}{1-\frac{1}{x}-\frac{6}{x^2}}$$

5. Simplify:

$$2+\frac{5}{x-\frac{1}{5}}$$

Matched Problems with Objectives Name ____________________

Equations With Rational Expressions

Section 5.5 Date ____________________

Directions Each problem below is similar to the example with the same number in your textbook. After reading through an example in your textbook, or watching one of the videos of that example on MathTV, try the matched problem to check your progress in this section. The shaded text is the learning objective associated with the matched problems that appear below the objective.

Objective A Solve equations containing rational expressions. (Problem Set exercises 1 – 46 are similar.)

1. Solve: $\frac{x}{3}+1=\frac{1}{2}$

2. Solve: $\frac{2}{a+5}=\frac{1}{3}$

SECTION 5.5

3. Solve: $\frac{x}{x+1}-\frac{1}{2}=\frac{-1}{x+1}$

4. Solve: $\frac{x}{x^2-9}-\frac{1}{x+3}=\frac{1}{4x-12}$

5. Solve: $1-\frac{2}{x}=\frac{8}{x^2}$

6. Solve: $\frac{y+1}{3(y+4)}=\frac{8}{(y+4)(y-4)}$

Matched Problems with Objectives
Equations With Rational Expressions
Section 5.5

Name ______________________

Date ______________________

Objective B Solve formulas containing rational expressions. (Problem Set exercises 49 – 54 and 63 and 64 are similar.)

7. Solve for y: $x = \dfrac{y+2}{y-1}$

8. Solve the formula $\dfrac{1}{a} = \dfrac{1}{x} + \dfrac{1}{b}$ for x.

Objective C Graph rational functions. (Problem Set exercises 55 – 62 are similar.)

9. Graph the rational function: $f(x) = \dfrac{2}{x-3}$

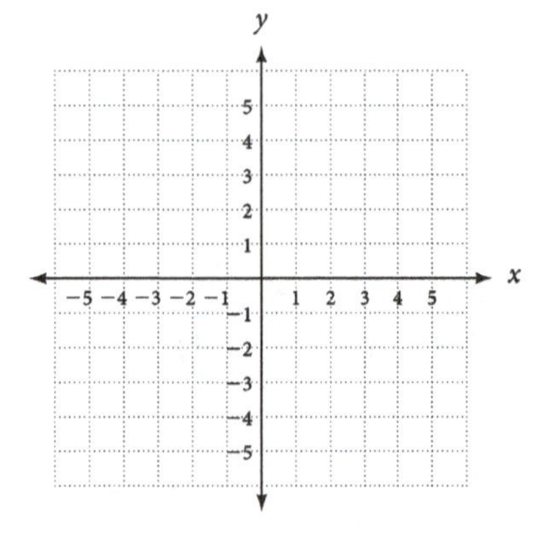

10. Graph the rational function: $f(x) = \dfrac{2}{x+3}$

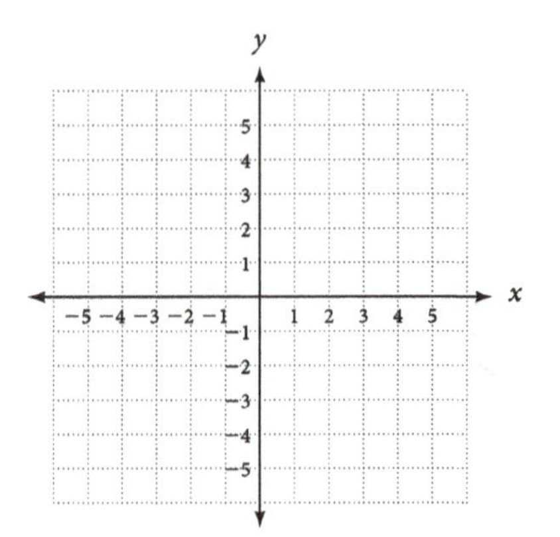

Matched Problems with Objectives Name ______________________

Applications

Section 5.6 Date ______________________

Directions Each problem below is similar to the example with the same number in your textbook. After reading through an example in your textbook, or watching one of the videos of that example on MathTV, try the matched problem to check your progress in this section. The shaded text is the learning objective associated with the matched problems that appear below the objective.

Objective A Solve application problems using equations containing rational expressions. (Problem Set exercises 1 – 28 are similar.)

1. One number is 3 times another. The sum of their reciprocals is $\frac{4}{3}$. Find the numbers.

2. A boat can travel at 15 miles per hour in still water. If it takes the same amount of time for the boat to travel 2 miles downstream as it does to travel 1 mile upstream, find the speed of the current.

3. The current of a river is 2 miles per hour. It takes a motorboat a total of 3 hours to travel 8 miles upstream and return 8 miles downstream. What is the speed of the boat in still water?

4. An inlet pipe can fill a pool in 10 hours, and a drain can empty it in 12 hours. If the pool is empty and both the inlet pipe and the drain are open, how long will it take to fill the pool?

Matched Problems with Objectives Name ____________________

Applications

Section 5.6 Date ____________________

Objective B Graph rational functions that have vertical and horizontal asymptotes. (Problem Set exercises 29 – 34 are similar.)

5. Graph the rational function $y = \frac{x-3}{x-1}$.

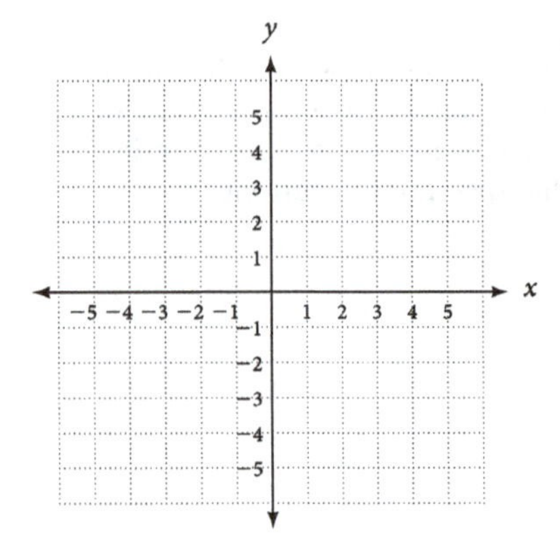

SECTION 5.6

Matched Problems with Objectives Name ______________________

Division of Polynomials

Section 5.7 Date ______________________

Directions Each problem below is similar to the example with the same number in your textbook. After reading through an example in your textbook, or watching one of the videos of that example on MathTV, try the matched problem to check your progress in this section. The shaded text is the learning objective associated with the matched problems that appear below the objective.

Objective A Divide a polynomial by a monomial. (Problem Set exercises 1 – 12 are similar.)

1. Divide: $\dfrac{12x^4 - 18x^3 + 24x^2}{6x}$

2. Divide and write the result with positive exponents: $\dfrac{27x^4y^7 - 81x^5y^3}{-9x^3y^2}$

SECTION 5.7

3. Divide and write the result with positive exponents: $\dfrac{12a^5 + 8a^4 + 16a^3 + 4a^2}{8a^4}$

Objective B Divide a polynomial by a polynomial. (Problem Set exercises 13 – 48 are similar.)

4. Divide: $\dfrac{2x^2 - 5xy + 3y^2}{x - y}$

5. Divide: $35\overline{)7,546}$

Matched Problems with Objectives Name ____________________

Division of Polynomials

Section 5.7 Date ____________________

6. Divide: $\dfrac{3x^2-8x-1}{x-3}$

7. Divide: $x-2\overline{)3x^3+3x+1}$

SECTION 5.7

8. Divide: $\dfrac{x^2-6xy-7y^2}{x+y}$

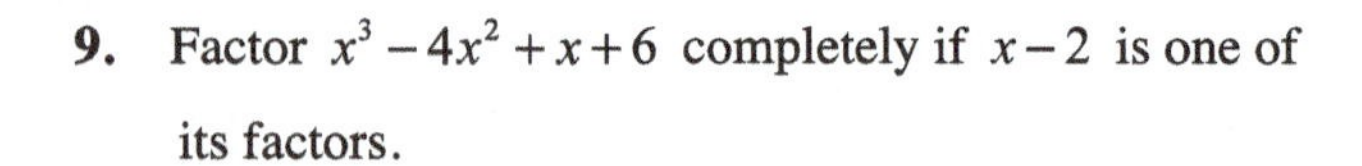

9. Factor x^3-4x^2+x+6 completely if $x-2$ is one of its factors.

Matched Problems with Objectives Name ____________________

Rational Exponents

Section 6.1 Date ____________________

Directions Each problem below is similar to the example with the same number in your textbook. After reading through an example in your textbook, or watching one of the videos of that example on MathTV, try the matched problem to check your progress in this section. The shaded text is the learning objective associated with the matched problems that appear below the objective.

Objective A Simplify radical expressions using the definition of roots. (Problem Set exercises 1 – 26 are similar.)

1. Give the two square roots of 36.

2. Simplify this expression, if possible: $\sqrt[3]{-64}$

3. Simplify this expression, if possible: $\sqrt{-25}$

4. Simplify this expression, if possible: $-\sqrt{4}$

5. Simplify this expression, if possible: $\sqrt[5]{-1}$

6. Simplify this expression, if possible: $\sqrt[4]{-16}$

7. Assume all variables represent nonnegative numbers and simplify this expression as much as possible: $\sqrt{81a^4b^8}$

8. Assume all variables represent nonnegative numbers and simplify this expression as much as possible: $\sqrt[3]{8x^3y^9}$

9. Assume all variables represent nonnegative numbers and simplify this expression as much as possible: $\sqrt[4]{81a^4b^8}$

Objective B Simplify expressions with rational exponents. (Problem Set exercises 27 – 78 are similar.)

10. Write this expression as a root and then simplify, if possible: $9^{\frac{1}{2}}$

11. Write this expression as a root and then simplify, if possible: $27^{\frac{1}{3}}$

12. Write this expression as a root and then simplify, if possible: $-49^{\frac{1}{2}}$

13. Write this expression as a root and then simplify, if possible: $(-49)^{\frac{1}{2}}$

Matched Problems with Objectives
Rational Exponents
Section 6.1

Name ______________________

Date ______________________

14. Write this expression as a root and then simplify, if possible: $\left(\frac{16}{25}\right)^{\frac{1}{2}}$

15. Write this radical with a rational exponent and then simplify: $\sqrt[3]{8x^3y^9}$

16. Write this radical with a rational exponent and then simplify: $\sqrt[4]{81a^4b^8}$

17. Simplify as much as possible: $9^{\frac{3}{2}}$

18. Simplify as much as possible: $16^{\frac{3}{4}}$

19. Simplify as much as possible: $8^{-\frac{2}{3}}$

20. Simplify as much as possible: $\left(\frac{16}{81}\right)^{-\frac{3}{4}}$

21. Assume all variables represent positive quantities and simplify as much as possible: $x^{\frac{1}{2}} \cdot x^{\frac{1}{4}}$

22. Assume all variables represent positive quantities and simplify as much as possible: $\left(y^{\frac{3}{5}}\right)^{\frac{5}{6}}$

23. Assume all variables represent positive quantities and simplify as much as possible: $\frac{z^{\frac{3}{4}}}{z^{\frac{2}{3}}}$

24. Assume all variables represent positive quantities and simplify as much as possible: $\left(\frac{a^{-\frac{1}{2}}}{b^{\frac{1}{4}}}\right)^8$

25. Assume all variables represent positive quantities and simplify as much as possible: $\frac{\left(x^{\frac{1}{3}}y^{-3}\right)^6}{x^4y^{10}}$

Matched Problems with Objectives Name ______________________

Simplified Form for Radicals

Section 6.2 Date ______________________

Directions Each problem below is similar to the example with the same number in your textbook. After reading through an example in your textbook, or watching one of the videos of that example on MathTV, try the matched problem to check your progress in this section. The shaded text is the learning objective associated with the matched problems that appear below the objective.

Objective A Write radical expressions in simplified form. (Problem Set exercises 1 – 40 and 71 – 82 are similar.)

1. Write $\sqrt{18}$ in simplified form.

2. Write $\sqrt{50x^2y^3}$ in simplified form. Assume $x, y \geq 0$.

3. Write $\sqrt[3]{54a^4b^3}$ in simplified form.

4. Write $\sqrt{75x^5y^8}$ in simplified form.

5. Write $\sqrt[4]{48a^8b^5c^4}$ in simplified form.

6. Write $\sqrt{\dfrac{5}{9}}$ in simplified form.

Objective B Rationalize a denominator that contains only one term. (Problem Set exercises 41 – 70 are similar.)

7. Write $\sqrt{\dfrac{2}{3}}$ in simplified form.

8. Rationalize the denominator: $\dfrac{5}{\sqrt{2}}$

Matched Problems with Objectives Name ______________________

Simplified Form for Radicals

Section 6.2 Date ______________________

9. Rationalize the denominator: $\dfrac{3\sqrt{5x}}{\sqrt{2y}}$

10. Rationalize the denominator: $\dfrac{5}{\sqrt[3]{9}}$

11. Write $\sqrt{\dfrac{48x^3y^4}{7z}}$ in simplified form.

12. Simplify this expression. Do not assume the variables represent positive numbers. $\sqrt{16x^2}$

SECTION 6.2

13. Simplify this expression. Do not assume the variables represent positive numbers. $\sqrt{25x^3}$

14. Simplify this expression. Do not assume the variables represent positive numbers. $\sqrt{x^2+10x+25}$

15. Simplify this expression. Do not assume the variables represent positive numbers. $\sqrt{2x^3+7x^2}$

16. Simplify: $\sqrt[3]{(-3)^3}$

17. Simplify: $\sqrt[3]{(-1)^3}$

Matched Problems with Objectives

Addition and Subtraction of Radical Expressions

Section 6.3

Name ____________________

Date ____________________

Directions Each problem below is similar to the example with the same number in your textbook. After reading through an example in your textbook, or watching one of the videos of that example on MathTV, try the matched problem to check your progress in this section. The shaded text is the learning objective associated with the matched problems that appear below the objective.

Objective A Add and subtract radicals. (Problem Set exercises 1 – 46 are similar.)

1. Combine: $3\sqrt{5}-2\sqrt{5}+4\sqrt{5}$

2. Combine: $4\sqrt{50}+3\sqrt{8}$

3. Assume $x, y \geq 0$ and combine:
$4\sqrt{18x^2y}-3x\sqrt{50y}$

4. Assume $a, b \geq 0$ and combine:
$2\sqrt[3]{27a^2b^4}+3b\sqrt[3]{125a^2b}$

SECTION 6.3

5. Combine: $\dfrac{\sqrt{5}}{3}+\dfrac{1}{\sqrt{5}}$

Objective B Construct golden rectangles from squares. (Problem Set exercises 47 – 52 are similar.)

6. Construct a golden rectangle from a square of side 6. Then show that the ratio of the length to the width is the golden ratio.

Matched Problems with Objectives

Name ______________________

Multiplication and Division of Radical Expressions

Section 6.4

Date ______________________

Directions Each problem below is similar to the example with the same number in your textbook. After reading through an example in your textbook, or watching one of the videos of that example on MathTV, try the matched problem to check your progress in this section. The shaded text is the learning objective associated with the matched problems that appear below the objective.

Objective A Multiply expressions containing radicals. (Problem Set exercises 1 – 38 are similar.)

1. Multiply: $(7\sqrt{3})(5\sqrt{11})$

2. Multiply: $\sqrt{2}(3\sqrt{5}-4\sqrt{2})$

3. Multiply: $(\sqrt{2}+\sqrt{7})(\sqrt{2}-3\sqrt{7})$

4. Expand and simplify: $(\sqrt{x}+5)^2$

5. Expand and simplify: $(5\sqrt{a}-3\sqrt{b})^2$

6. Expand and simplify: $(\sqrt{x+3}-1)^2$

7. Multiply: $(\sqrt{5}+\sqrt{3})(\sqrt{5}-\sqrt{3})$

Matched Problems with Objectives Name ____________________

Multiplication and Division of Radical Expressions

Section 6.4 Date ____________________

Objective B Rationalize a denominator containing two terms. (Problem Set exercises 39 – 60 are similar.)

8. Divide by rationalizing the denominator: $\frac{3}{\sqrt{7}-\sqrt{3}}$

9. Rationalize the denominator: $\frac{\sqrt{10}-3}{\sqrt{10}+3}$

10. If side *AC* in Figure 1 in your textbook was 4 instead of 2, show that the smaller rectangle *BDEF* is also a golden rectangle by finding the ratio of its length to its width.

Matched Problems with Objectives Name ____________________

Equations Involving Radicals

Section 6.5 Date ____________________

Directions Each problem below is similar to the example with the same number in your textbook. After reading through an example in your textbook, or watching one of the videos of that example on MathTV, try the matched problem to check your progress in this section. The shaded text is the learning objective associated with the matched problems that appear below the objective.

Objective A Solve equations containing radicals. (Problem Set exercises 1 – 48 are similar.)

1. Solve: $\sqrt{2x+4}=4$

2. Solve: $\sqrt{7x-3}=-5$

3. Solve: $\sqrt{4x+5}+2=7$

4. Solve: $t-6=\sqrt{t-4}$

5. Solve: $\sqrt{x-9}=\sqrt{x}-3$

6. Solve: $\sqrt{x+4}=2-\sqrt{3x}$

Matched Problems with Objectives Name ____________________

Equations Involving Radicals

Section 6.5 Date ____________________

7. Solve: $\sqrt{x+2} = \sqrt{x+3} - 1$

8. Solve: $\sqrt[3]{3x-7} = 2$

Objective B Graph simple square root and cube root equations in two variables. (Problem Set exercises 53 – 64 are similar.)

9. Graph each equation:

$y = \sqrt{x} + 3$

$y = \sqrt{x+3}$

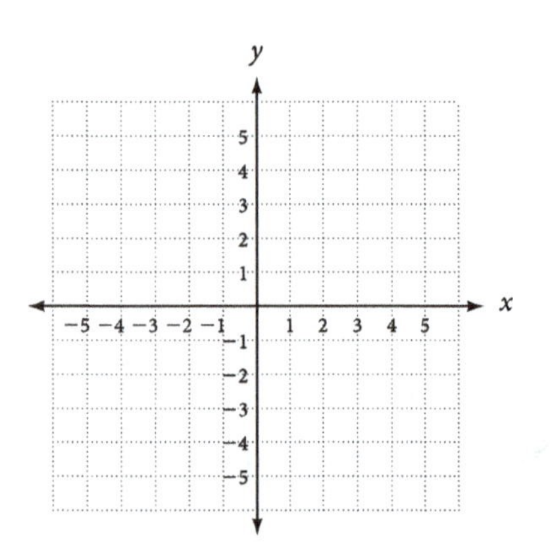

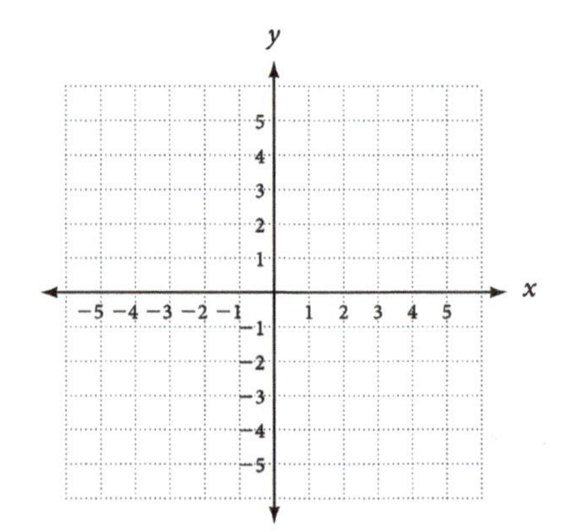

Matched Problems with Objectives

Complex Numbers

Section 6.6

Name ______________________

Date ______________________

Directions Each problem below is similar to the example with the same number in your textbook. After reading through an example in your textbook, or watching one of the videos of that example on MathTV, try the matched problem to check your progress in this section. The shaded text is the learning objective associated with the matched problems that appear below the objective.

Objective A Simplify square roots of negative numbers. (Problem Set exercises 1 – 8 are similar.)

1. Write this square root in terms of the number i:
$\sqrt{-36}$

2. Write this square root in terms of the number i:
$-\sqrt{-64}$

3. Write this square root in terms of the number i:
$\sqrt{-18}$

4. Write this square root in terms of the number i:
$-\sqrt{-19}$

Objective B Simplify powers of i. (Problem Set exercises 9 – 14are similar.)

5. Simplify: i^{20}

6. Simplify: i^{23}

7. Simplify: i^{50}

Objective C Solve for unknown variables by equating real parts and equating imaginary parts of two complex numbers. (Problem Set exercises 15 – 24 are similar.)

8. Find x and y if $4x+7i=8-14yi$.

9. Find x and y if $(2x-1)+9i=5+(4y+1)i$.

Matched Problems with Objectives Name ____________________

Complex Numbers

Section 6.6 Date ____________________

Objective D Add and subtract complex numbers. (Problem Set exercises 25 – 40 are similar.)

10. Add: $(2+6i)+(3-4i)$

11. Subtract: $(6+5i)-(4+3i)$

12. Subtract: $(7-i)-(8-2i)$

Objective E Multiply complex numbers. (Problem Set exercises 41 – 66 are similar.)

SECTION 6.6

13. Multiply: $(2+3i)(1-4i)$

14. Multiply: $(-3i)(2+3i)$

15. Expand: $(2+4i)^2$

16. Multiply: $(3+5i)(3-5i)$

Objective F Divide complex numbers. (Problem Set exercises 67 – 78 are similar.)

17. Divide: $\dfrac{3+2i}{2-5i}$

18. Divide: $\dfrac{3+2i}{i}$

Matched Problems with Objectives Name ____________

Completing the Square

Section 7.1 Date ____________

Directions Each problem below is similar to the example with the same number in your textbook. After reading through an example in your textbook, or watching one of the videos of that example on MathTV, try the matched problem to check your progress in this section. The shaded text is the learning objective associated with the matched problems that appear below the objective.

Objective A Solve quadratic equations by taking the square root of both sides. (Problem Set exercises 1 – 20 are similar.)

1. Solve: $(3x+2)^2=16$

2. Solve: $(4x-3)^2=-50$

3. Solve: $x^2+10x+25=20$

Objective B Solve quadratic equations by completing the square. (Problem Set exercises 21 – 56 are similar.)

4. Solve by completing the square: $x^2+3x-4=0$

5. Solve by completing the square: $5x^2-3x+2=0$

Matched Problems with Objectives Name ______________________

Completing the Square

Section 7.1 Date ______________________

Objective C Use quadratic equations to solve for missing parts of right triangles. (Problem Set exercises 67 – 76 are similar.)

6. If the shortest side in a $30° - 60° - 90°$ triangle is 2 inches long, find the lengths of the other two sides.

7. Table 1 in the introduction to this chapter gives the vertical rise of the Lookout Double chair lift as 960 feet and the length of the chair lift as 4,330 feet. To the nearest foot, find the horizontal distance covered by a person riding this lift.

Matched Problems with Objectives Name ____________________

The Quadratic Formula

Section 7.2 Date ____________________

Directions Each problem below is similar to the example with the same number in your textbook. After reading through an example in your textbook, or watching one of the videos of that example on MathTV, try the matched problem to check your progress in this section. The shaded text is the learning objective associated with the matched problems that appear below the objective.

Objective A Solve quadratic equations by the quadratic formula. (Problem Set exercises 1 – 48 are similar.)

1. Solve using the quadratic formula: $6x^2+7x+2=0$

2. Solve using the quadratic formula: $\frac{x^2}{2}+x=\frac{1}{3}$

SECTION 7.2

3. Solve using the quadratic formula: $\frac{1}{x+4}-\frac{1}{x}=\frac{1}{2}$

4. Solve using the quadratic formula: $8t^3-27=0$

Matched Problems with Objectives Name ____________________

The Quadratic Formula

Section 7.2 Date ____________________

Objective B Solve application problems using quadratic equations. (Problem Set exercises 57 – 60 are similar.)

5. An object thrown upward with an initial velocity of 32 feet per second rises and falls according to the equation $s = 32t - 16t^2$ where s is the height of the object above the ground at any time t. At what times will the object be 12 feet above the ground?

6. A company produces and sells copies of an accounting program for home computers. The total weekly profit (in dollars) of x copies of the program is $P(x) = -500 + 27x - 0.1x^2$. How many copies must be sold for its weekly profit to be $1,320?

SECTION 7.2

Matched Problems with Objectives Name ____________________

Additional Items Involving Solutions to Equations

Section 7.3 Date ____________________

Directions Each problem below is similar to the example with the same number in your textbook. After reading through an example in your textbook, or watching one of the videos of that example on MathTV, try the matched problem to check your progress in this section. The shaded text is the learning objective associated with the matched problems that appear below the objective.

Objective A Find the number and kind of solutions to a quadratic equation by using the discriminant. (Problem Set exercises 1 – 12 are similar.)

1. Give the number and kind of solution: $x^2-3x-28=0$

2. Give the number and kind of solution: $x^2-6x+9=0$

3. Give the number and kind of solution: $3x^2-2x+4=0$

4. Give the number and kind of solution: $x^2+1=4x$

Objective B Find an unknown constant in a quadratic equation so that there is exactly one solution. (Problem Set exercises 13 – 22 are similar.)

5. Find k so that the equation $9x^2+kx=-4$ has exactly one rational solution.

Matched Problems with Objectives

Additional Items Involving Solutions to Equations

Section 7.3

Name ______________________

Date ______________________

Objective C Find an equation from its solutions. (Problem Set exercises 23 – 52 are similar.)

6. Find an equation that has solutions $t = -2$, $t = 2$, and $t = 3$.

7. Find an equation that has solutions $x = -\frac{3}{4}$ and $x = \frac{1}{5}$.

SECTION 7.3

Matched Problems with Objectives Name ____________________

More Equations

Section 7.4 Date ____________________

Directions Each problem below is similar to the example with the same number in your textbook. After reading through an example in your textbook, or watching one of the videos of that example on MathTV, try the matched problem to check your progress in this section. The shaded text is the learning objective associated with the matched problems that appear below the objective.

Objective A Solve equations that are reducible to a quadratic equation. (Problem Set exercises 1 – 32 are similar.)

1. Solve: $(x-2)^2-3(x-2)-10=0$

2. Solve: $6x^4-13x^2=5$

3. Solve: $x-\sqrt{x}-12=0$

SECTION 7.4

Objective B Solve application problems using equations quadratic in form. (Problem Set exercises 33 – 38 are similar.)

4. If an object is tossed into the air with an upward velocity of 10 feet per second from the top of a building h feet high, the time it takes for the object to hit the ground below is $16t^2-10t-h=0$. Solve this formula for t.

Matched Problems with Objectives Name ______________________

Graphing Parabolas

Section 7.5 Date ______________________

Directions Each problem below is similar to the example with the same number in your textbook. After reading through an example in your textbook, or watching one of the videos of that example on MathTV, try the matched problem to check your progress in this section. The shaded text is the learning objective associated with the matched problems that appear below the objective.

Objective A Graph a parabola. (Problem Set exercises 1 – 38 are similar.)

1. Graph: $y = x^2 - 2x - 3$

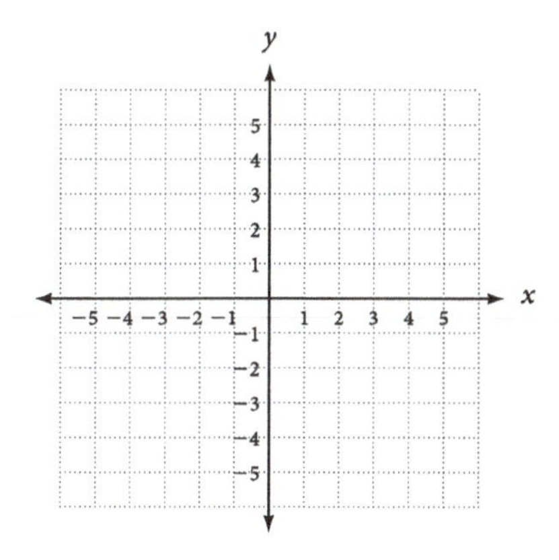

2. Graph: $y = -x^2 + 2x + 8$

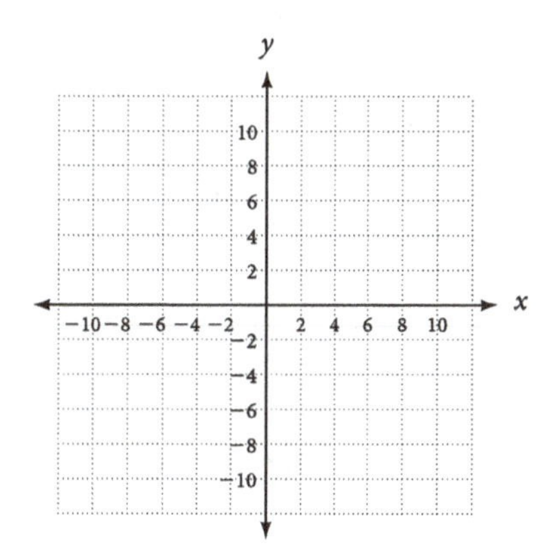

SECTION 7.5

3. Graph: $y = 2x^2 - 4x + 1$

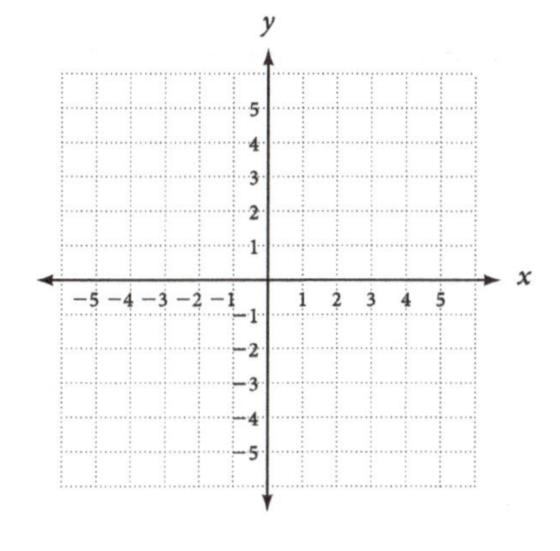

4. Graph: $y = -x^2 + 4x - 5$

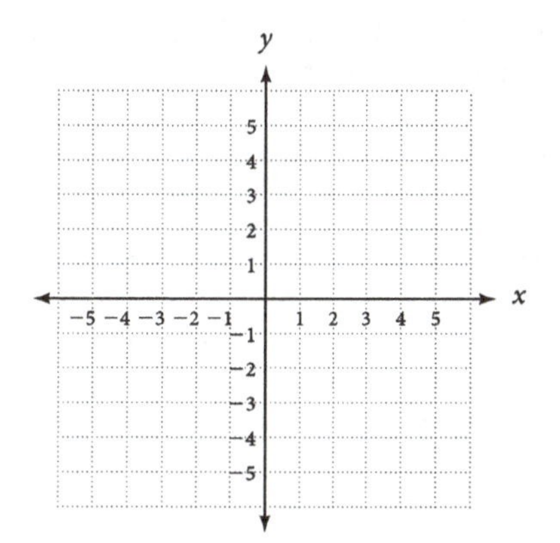

Matched Problems with Objectives Name ______________________

Graphing Parabolas

Section 7.5 Date ______________________

Objective B Solve application problems using information from a graph. (Problem Set exercises 39 – 46 are similar.)

5. Find the largest value of y if $y = -0.01x^2 + 12x - 400$.

6. An art supply store finds that they can sell x sketch pads each week at p dollars each according to the equation $x = 800 - 200p$. Graph the revenue equation $R = xp$. Then use the graph to find the price that will bring in the maximum revenue. Also, find the maximum revenue.

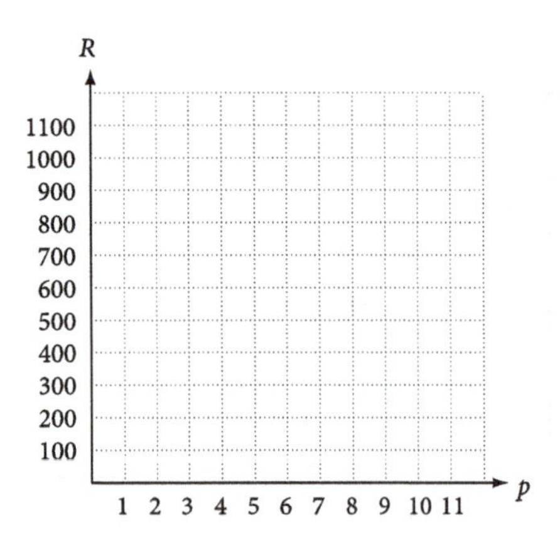

SECTION 7.5

Objective C Find an equation from its graph. (Problem Set exercises 47 – 48 are similar.)

7. Assume David Smith, Jr., The Bullet, was shot from a cannon and he reached a maximum height of 80 feet before landing in a net 160 feet from the cannon. Sketch the graph of his path, and then find the equation of the graph.

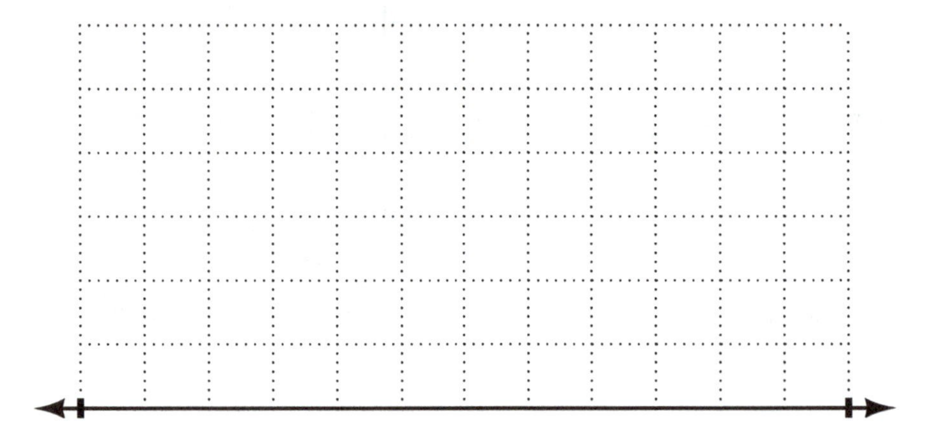

Matched Problems with Objectives Name ____________________

Quadratic Inequalities

Section 7.6 Date ____________________

Directions Each problem below is similar to the example with the same number in your textbook. After reading through an example in your textbook, or watching one of the videos of that example on MathTV, try the matched problem to check your progress in this section. The shaded text is the learning objective associated with the matched problems that appear below the objective.

Objective A Solve quadratic inequalities and graph the solution set. (Problem Set exercises 1 – 22 are similar.)

1. Solve for x: $x^2 + 2x - 15 \le 0$

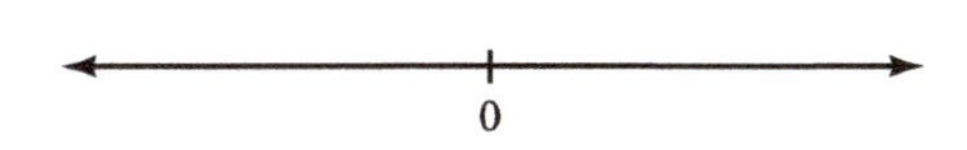

2. Solve for x: $2x^2 + 3x > 2$

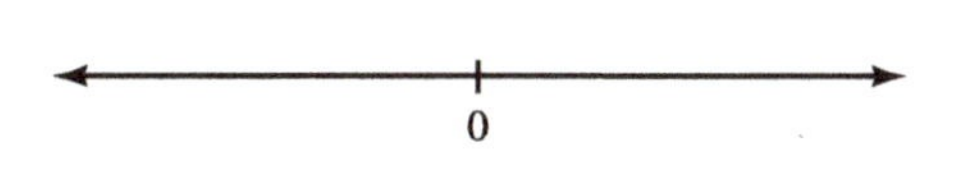

SECTION 7.6

3. Solve for x: $x^2 + 10x + 25 < 0$

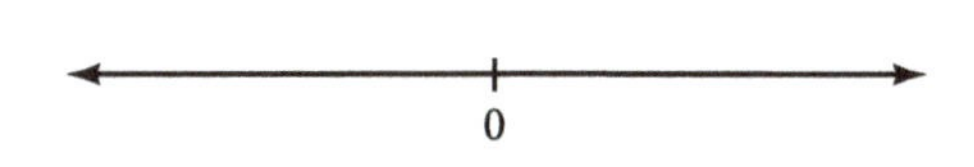

Matched Problems with Objectives Name ____________________

Quadratic Inequalities

Section 7.6 Date ____________________

Objective B Solve a rational inequality and graph the solution set. (Problem Set exercises 23 – 34 are similar.)

4. Solve for x: $\dfrac{x+3}{x-1} \le 0$

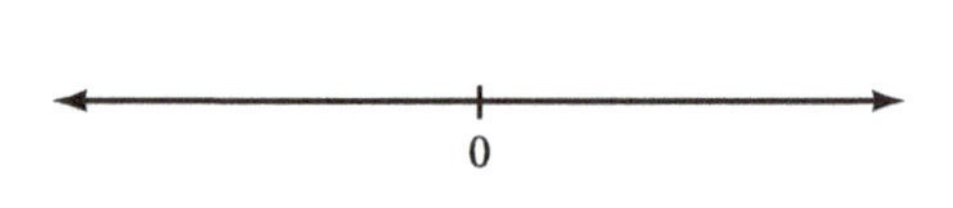

5. Solve for x: $\dfrac{3}{x+1} - \dfrac{2}{x-4} > 0$

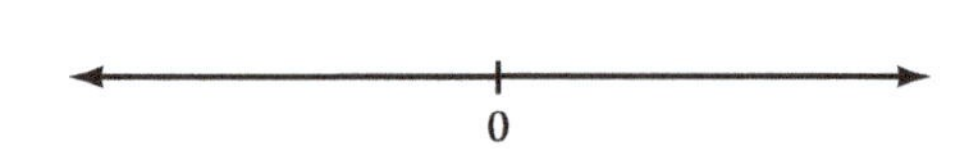

Matched Problems with Objectives
Exponential Functions
Section 8.1

Name ______________________

Date ______________________

Directions Each problem below is similar to the example with the same number in your textbook. After reading through an example in your textbook, or watching one of the videos of that example on MathTV, try the matched problem to check your progress in this section. The shaded text is the learning objective associated with the matched problems that appear below the objective.

Objective A Find function values for exponential functions. (Problem Set exercises 1 – 8 are similar.)

1. If $f(x) = 4^x$, find the following:

a. $f(0)$ **d.** $f(3)$

b. $f(1)$ **e.** $f(-1)$

c. $f(2)$ **f.** $f(-2)$

2. Let $A(t) = 1200 \cdot 2^{-\frac{t}{8}}$. Find $A(12)$ and $A(24)$.

Objective B Graph exponential functions. (Problem Set exercises 25 – 28 are similar.)

3. Sketch the graph of $y = 3^x$.

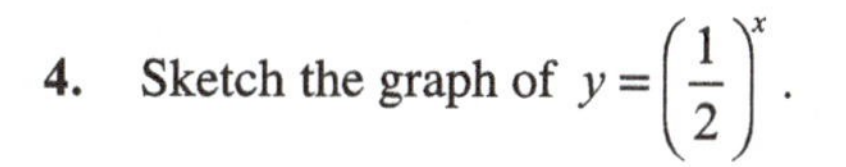

4. Sketch the graph of $y = \left(\frac{1}{2}\right)^x$.

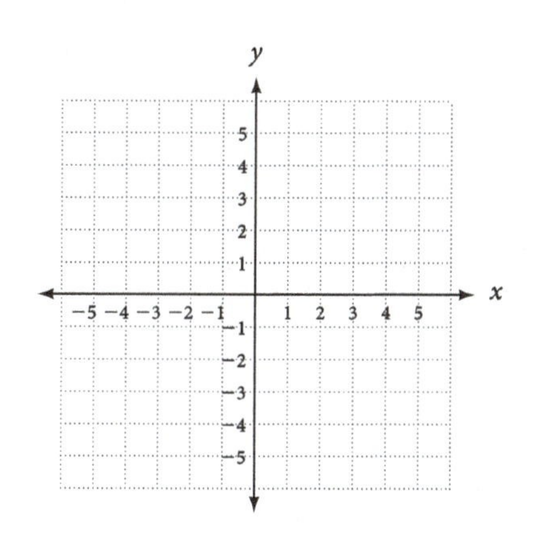

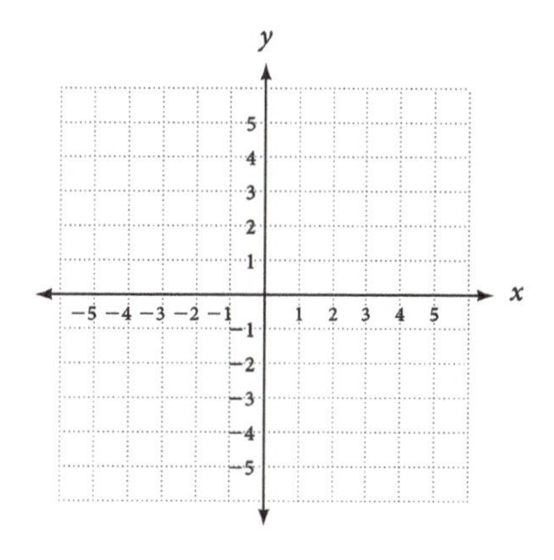

MathTV.com

Matched Problems with Objectives Name ______________________

Exponential Functions

Section 8.1 Date ______________________

Objective C Solve applications involving exponential growth and decay. (Problem Set exercises 29 – 46 are similar.)

5. Suppose you deposit $600 in an account with an annual interest rate of 6% compounded quarterly.

a. Find an equation that gives the amount of money in the account after t years.

b. Find the amount in the account after 5 years.

c. Find the number of years it will take to accumulate $1,000.

6. Suppose you deposit $600 in an account with an annual interest rate of 6% compounded continuously. How much money is in the account after 5 years?

Matched Problems with Objectives Name ______________________

The Inverse of a Function Date ______________________

Section 8.2

Directions Each problem below is similar to the example with the same number in your textbook. After reading through an example in your textbook, or watching one of the videos of that example on MathTV, try the matched problem to check your progress in this section. The shaded text is the learning objective associated with the matched problems that appear below the objective.

Objective A Find the equation of the inverse of a function. (Problem Set exercises 1 – 18 are similar.)

1. If $f(x) = 4x + 1$, find the equation for its inverse.

2. Find the inverse of $y = x^2 + 1$. Then graph the function and its inverse on the same axes.

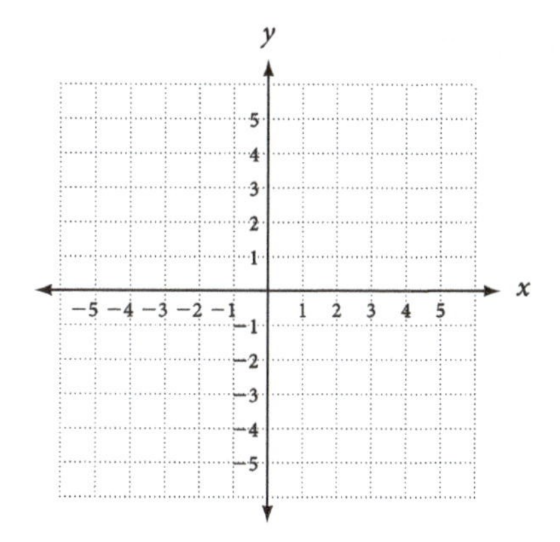

SECTION 8.2

Matched Problems with Objectives Name ____________________

The Inverse of a Function Date ____________________

Section 8.2

3. If $g(x)=\dfrac{x-3}{x+1}$, find the equation for its inverse.

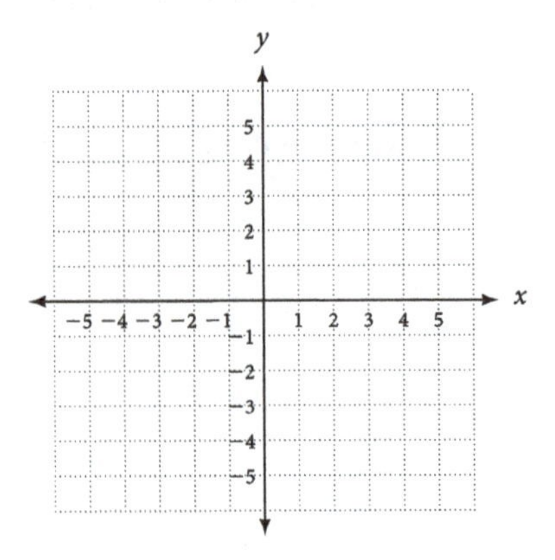

Objective B Graph a function and its inverse. (Problem Set exercises 19 –40 are similar.)

SECTION 8.2

4. Graph the function $y=3^x$ and its inverse $x=3^y$ on the same axes.

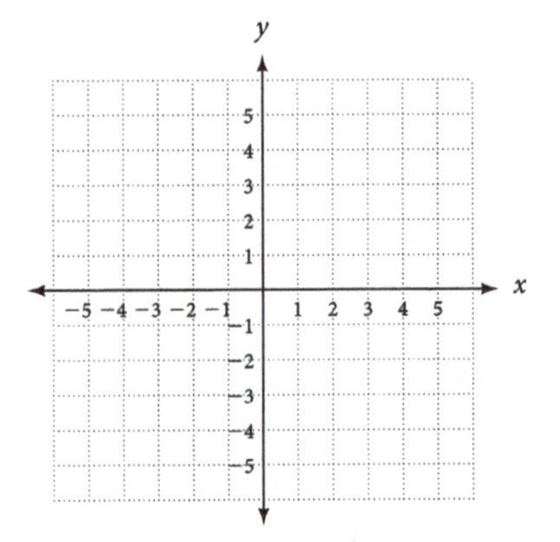

Matched Problems with Objectives Name ____________________

Logarithms are Exponents

Section 8.3 Date ____________________

Directions Each problem below is similar to the example with the same number in your textbook. After reading through an example in your textbook, or watching one of the videos of that example on MathTV, try the matched problem to check your progress in this section. The shaded text is the learning objective associated with the matched problems that appear below the objective.

Objective A Convert between logarithmic form and exponential form. (Problem Set exercises 1 – 24 are similar.)

1. Convert to exponential form and then solve for x: $\log_2 x = 3$

Objective B Use the definition of logarithms to solve simple logarithmic equations. (Problem Set exercises 25 – 46 are similar.)

2. Solve: $\log_x 5 = 2$

3. Solve: $\log_9 27 = x$

Matched Problems with Objectives Name ______________________

Logarithms are Exponents

Section 8.3 Date ______________________

Objective C Sketch the graph of a logarithmic function.. (Problem Set exercises 47 – 58 are similar.)

4. Graph the equation $y = \log_3 x$.

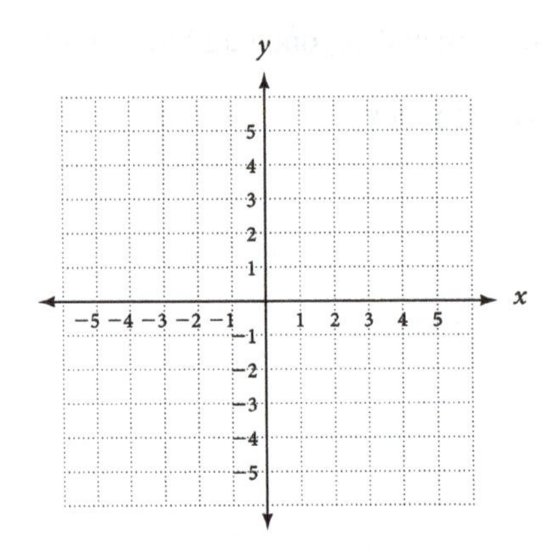

Objective D Simplify expressions involving logarithms. (Problem Set exercises 59 – 82 are similar.)

5. Simplify the following expressions:

a. $\log_3 27$

b. $\log_{10} 1000$

c. $\log_6 6$

d. $\log_3 1$

e. $\log_2 (\log_8 8)$

Objective E Solve applications involving logarithmic equations.. (Problem Set exercises 83 – 90 are similar.)

6. If an earthquake has a magnitude of $M = 6$ on the Richter scale, what can you say about the size of its shock wave?

Matched Problems with Objectives Name ____________________

Properties of Logarithms

Section 8.4 Date ____________________

Directions Each problem below is similar to the example with the same number in your textbook. After reading through an example in your textbook, or watching one of the videos of that example on MathTV, try the matched problem to check your progress in this section. The shaded text is the learning objective associated with the matched problems that appear below the objective.

Objective A Use the properties of logarithms to convert between expanded for and single logarithms. (Problem Set exercises 1 – 42 are similar.)

1. Expand, using the properties of logarithms: $\log_3 \frac{5a}{b}$

2. Expand: $\log_{10} \frac{x^2}{\sqrt[3]{y}}$

3. Write as a single logarithm: $3\log_4 x + \log_4 y - 2\log_4 z$

Objective B Use the properties of logarithms to solve equations that contain logarithms. (Problem Set exercises 43 – 66 are similar.)

4. Solve for x: $\log_2(x+3) + \log_2 x = 2$

Matched Problems with Objectives

Name ______________________

Common Logarithms and Natural Logarithms

Section 8.5

Date ______________________

Directions Each problem below is similar to the example with the same number in your textbook. After reading through an example in your textbook, or watching one of the videos of that example on MathTV, try the matched problem to check your progress in this section. The shaded text is the learning objective associated with the matched problems that appear below the objective.

Objective A Use a calculator to find common logarithms. (Problem Set exercises 1 – 20 are similar.)

1. Use a calculator to find $\log(27,600)$.

2. Use a calculator to find $\log(0.00391)$.

3. Use a calculator to find $\log(0.00952)$.

Objective B Use a calculator to find a number given its common logarithm. (Problem Set exercises 21 – 46 are similar.)

4. Find x if $\log x = 3.9786$.

5. Find x if $\log x = -1.5901$.

Objective C Solve applications that involve logarithms. (Problem Set exercises 77 –96 are similar.)

6. Find T if an earthquake measures 5.5 on the Richter scale.

7. Find the concentration of the hydrogen ion in a can of cola if the pH is 4.1.

Matched Problems with Objectives Name ______________________

Common Logarithms and Natural Logarithms

Section 8.5 Date ______________________

8. The concentration of hydrogen ion in a sample of acid rain is 1.8×10^{-5}. Find the pH and round to the nearest tenth.

Objective D Simplify expressions containing natural logarithms. (Problem Set exercises 47 – 76 are similar.)

9. Simplify each expression:

a. $\ln e^2$

b. $\ln e^4$

c. $\ln e^{-2}$

d. $\ln e^x$

10. Expand $\ln Pe^{rt}$.

SECTION 8.5

11. If $\ln 5 = 1.6094$ and $\ln 7 = 1.9459$, find each of the following:

a. $\ln 35$

b. $\ln 0.2$

c. $\ln 49$

Matched Problems with Objectives Name ______________________

Exponential Equations and Change of Base

Section 8.6 Date ______________________

Directions Each problem below is similar to the example with the same number in your textbook. After reading through an example in your textbook, or watching one of the videos of that example on MathTV, try the matched problem to check your progress in this section. The shaded text is the learning objective associated with the matched problems that appear below the objective.

Objective A Solve exponential equations. (Problem Set exercises 1 – 30 are similar.)

1. Solve for x: $12^{x+2} = 20$

2. How long does it take for \$5,000 to double if it is invested in an account that pays 11% interest compounded once a year?

Objective B Use the change-of-base property. (Problem Set exercises 31 – 42 are similar.)

3. Find $\log_6 14$.

Objective C Solve application problems whose solutions are found by solving logarithmic or exponential equations. (Problem Set exercises 51 – 74 are similar.)

4. Suppose that the population in a small city is 32,000 in the beginning of 1994 and that the city council assumes that the population size t years later can be estimated by the equation, $P = 32{,}000e^{0.05t}$. Approximately when will the city have a population of 75,000?

Matched Problems with Objectives
More About Sequences
Section 9.1

Name ______________________
Date ______________________
Section ______________________

Directions Each problem below is similar to the example with the same number in your textbook. After reading through an example in your textbook, or watching one of the videos of that example on MathTV, try the matched problem to check your progress in this section. The shaded text is the learning objective associated with the matched problems that appear below the objective.

Objective A Write the terms of a sequence, given the general term. (Problem Set exercises 1 – 24 are similar.)

1. Find the first four terms of the sequence whose general term is given by $a_n = 3n - 2$.

2. Write the first four terms of the sequence defined by $a_n = \frac{3}{n+2}$.

3. Find the fifth and sixth terms of the sequence whose general term is given by $a_n = \frac{(-1)^{n+1}}{2n^2}$.

Objective B Write terms of a sequence given recursively. (Problem Set exercises 25 – 34 are similar.)

4. Write the first four terms of the sequence given recursively by $a_1 = 3$ and $a_n = 3a_{n-1} + 1$ for $n > 1$.

SECTION 9.1

Objective C Find the general term for a sequence. (Problem Set exercises 35 – 50 are similar.)

5. Find a formula for the *n*th term of the sequence

$3, 6, 9, 12, \ldots$

6. Find the general term of the sequence

$\frac{3}{4}, \frac{4}{5}, \frac{5}{6}, \frac{6}{7}, \ldots$

Matched Problems with Objectives
Series
Section 9.2

Name ______________________
Date ______________________
Section ______________________

Directions Each problem below is similar to the example with the same number in your textbook. After reading through an example in your textbook, or watching one of the videos of that example on MathTV, try the matched problem to check your progress in this section. The shaded text is the learning objective associated with the matched problems that appear below the objective.

Objective A Expand and simplify a series given by summation notation. (Problem Set exercises 1 – 28 are similar.)

1. Expand and simplify: $\sum_{i=1}^{5}(4i+2)$

2. Expand and simplify: $\sum_{i=2}^{5}\left(-\frac{1}{2}\right)^{i}$

3. Expand: $\sum_{i=2}^{5}(x^{i}+2)$

Objective B Write a series using summation notation. (Problem Set exercises 29 – 48 are similar.)

4. Write with summation notation: $3+6+9+12+15$

(There is more than one answer.)

5. Write with summation notation: $2+8+18+32$

(There is more than one answer.)

SECTION 9.2

6. Write with summation notation: $\frac{x+2}{x^2}+\frac{x+4}{x^4}+\frac{x+6}{x^6}+\frac{x+8}{x^8}$ (There is more than one answer.)

Matched Problems with Objectives
Arithmetic Sequences
Section 9.3

Name ______________________
Date ______________________
Section ______________________

Directions Each problem below is similar to the example with the same number in your textbook. After reading through an example in your textbook, or watching one of the videos of that example on MathTV, try the matched problem to check your progress in this section. The shaded text is the learning objective associated with the matched problems that appear below the objective.

Objective A Identify an arithmetic sequence (or arithmetic progression) and find its common difference. (Problem Set exercises 1 –10 are similar.)

1. Give the common difference for the arithmetic sequence: $5, 8, 11, 14, \ldots$

2. Give the common difference for the arithmetic sequence: $100, 96, 92, 88, \ldots$

3. Give the common difference for the arithmetic sequence: $\frac{5}{2}, 2, \frac{3}{2}, 1, \ldots$

Objective B Find the general term and the sum of an arithmetic sequence (or arithmetic progression) using the formulas. (Problem Set exercises 11 – 34 are similar.)

4. Find the general term of the arithmetic progression $5, 11, 17, 23, \ldots$

5. Find the general term of the arithmetic progression whose third term is 7 and whose eighth term is 22.

6. Find the sum of the first 10 terms of the arithmetic progression $-1, 4, 9, 14, 19, \ldots$

Matched Problems with Objectives
Geometric Sequences
Section 9.4

Name ______________________
Date ______________________
Section ______________________

Directions Each problem below is similar to the example with the same number in your textbook. After reading through an example in your textbook, or watching one of the videos of that example on MathTV, try the matched problem to check your progress in this section. The shaded text is the learning objective associated with the matched problems that appear below the objective.

Objective A Identify a geometric sequence (or geometric progression) and find its common ratio. (Problem Set exercises 1 – 10 are similar.)

1. Find the common ratio for the geometric progression $2, 6, 18, 54, \ldots$

2. Find the common ratio for the geometric progression $\sqrt{2}, 2, 2\sqrt{2}, 4, \ldots$

Objective B Find the general term and the sum of a geometric sequence (or geometric progression) using the given formulas. (Problem Set exercises 11 – 34 are similar.)

3. Find the general term for the geometric progression

$-6, 12, -24, 48, \ldots$

4. Find the tenth term of the geometric sequence

$4, 2, 1, \frac{1}{2}, \frac{1}{4}, \ldots$

Matched Problems with Objectives
Geometric Sequences
Section 9.4

Name ______________________
Date ______________________
Section ______________________

5. Find the general term for the geometric progression whose fourth term is 135 and whose seventh term is 3,645.

6. Find the sum of the first ten terms of the geometric sequence 3, 15, 75, 375, . . .

Objective C Find the sum of an infinite geometric series. (Problem Set exercises 35 – 50 are similar.)

7. Find the sum of the infinite geometric series

$$6-3+\frac{3}{2}-\frac{3}{4}+\ldots$$

8. Show that 0.66666… is the same as $\frac{2}{3}$.

SECTION 9.4

Matched Problems with Objectives
The Circle
Section 10.1

Name ______________________
Date ______________________
Section ______________________

Directions Each problem below is similar to the example with the same number in your textbook. After reading through an example in your textbook, or watching one of the videos of that example on MathTV, try the matched problem to check your progress in this section. The shaded text is the learning objective associated with the matched problems that appear below the objective.

Objective A Use the distance formula. (Problem Set exercises 1 – 14 are similar.)

1. Find the distance between $(-4,1)$ and $(2,5)$.

2. Find x if the distance from $(x,2)$ to $(3,-1)$ is $\sqrt{10}$.

Objective B Write the equation of a circle, given its center and radius. (Problem set exercises 15 – 22 are similar.)

3. Find the equation of the circle with center at $(4,-3)$ having a radius of 2.

4. Give the equation of the circle with radius 5 whose center is at the origin.

Matched Problems with Objectives
The Circle
Section 10.1

Name ______________________
Date ______________________
Section ______________________

Objective C Find the center and radius of a circle from its equation, and then sketch the graph. (Problem Set exercises 23 – 60 are similar.)

5. Find the center and radius of the circle whose equation is $(x-3)^2+(y-4)^2=9$ and then sketch the graph.

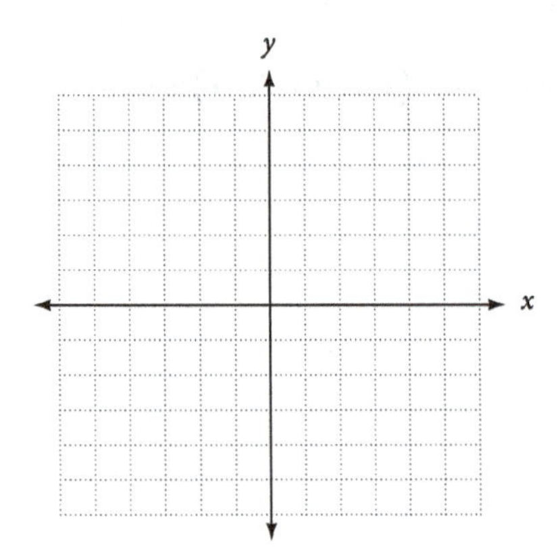

6. Sketch the graph of $x^2+y^2=25$.

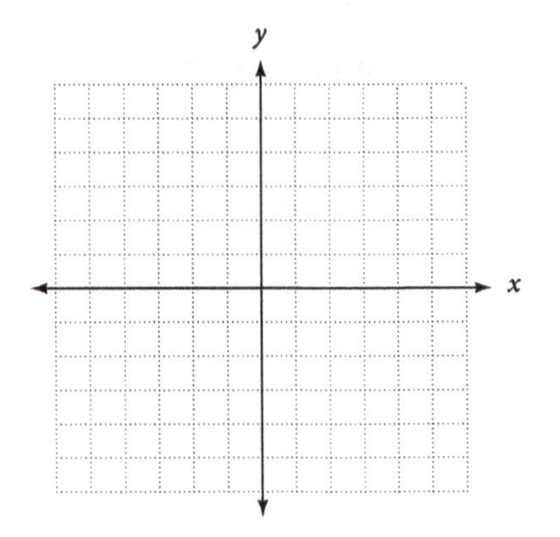

7. Sketch the graph of $x^2+y^2-6x+4y-3=0$.

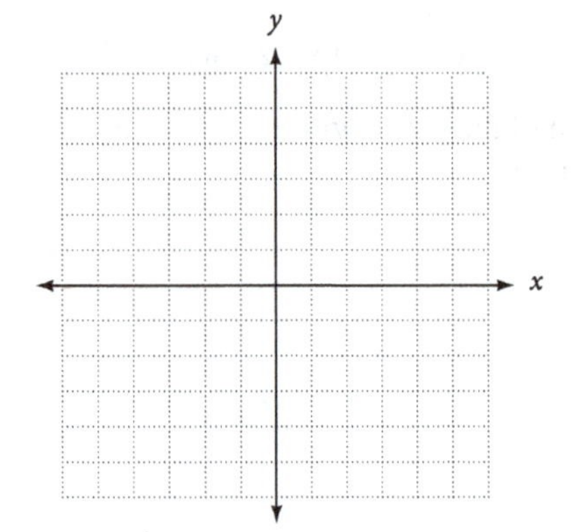

Matched Problems with Objectives
Ellipses and Hyperbolas
Section 10.2

Name ____________________
Date ____________________
Section ____________________

Directions Each problem below is similar to the example with the same number in your textbook. After reading through an example in your textbook, or watching one of the videos of that example on MathTV, try the matched problem to check your progress in this section. The shaded text is the learning objective associated with the matched problems that appear below the objective.

Objective A Graph an ellipse with center at the origin. (Problem Set exercises 1 – 10 are similar.)

1. Graph $25x^2 + 4y^2 = 100$.

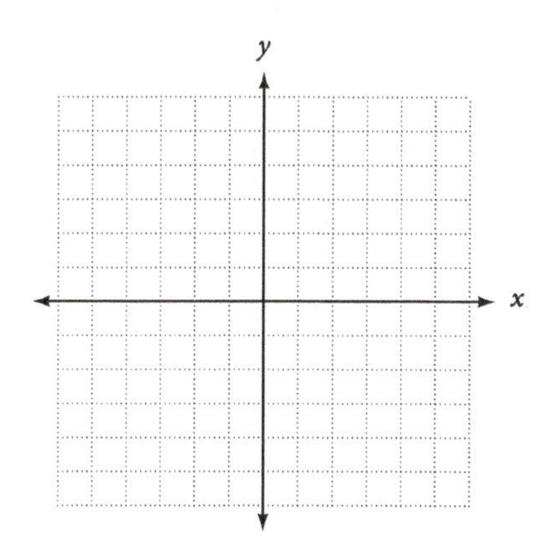

Objective B Graph a hyperbola with center at the origin. (Problem Set exercises 11 – 22 are similar.)

2. Graph $\frac{x^2}{25} - \frac{y^2}{9} = 1$

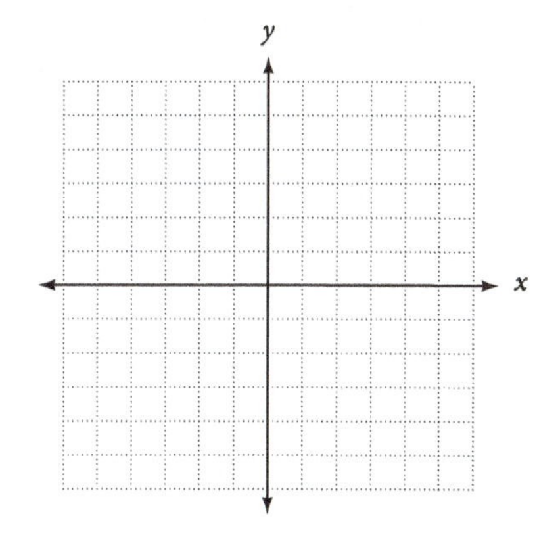

Matched Problems with Objectives
Ellipses and Hyperbolas
Section 10.2

Name ______________________
Date ______________________
Section ______________________

Objective C Graph an ellipse with center at (h, k). (Problem Set exercises 29 – 34 are similar.)

3. Graph $\frac{(x-3)^2}{16}+\frac{(y+1)^2}{4}=1$

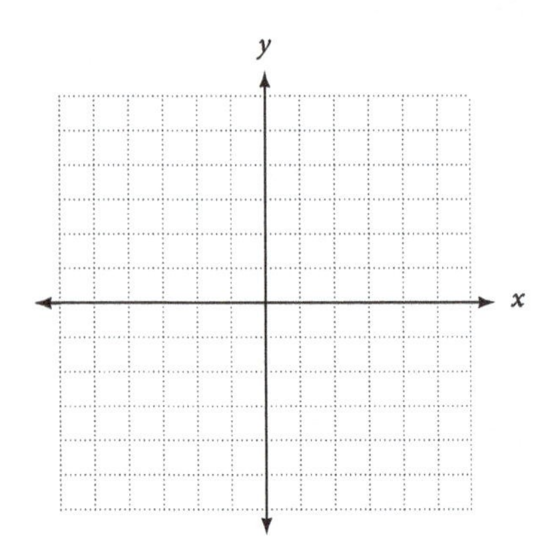

Objective D Graph a hyperbola with center at (h, k). (Problem Set exercises 35 – 40 are similar.)

4. Graph $\frac{(x+2)^2}{9}-\frac{(y-3)^2}{16}=1$

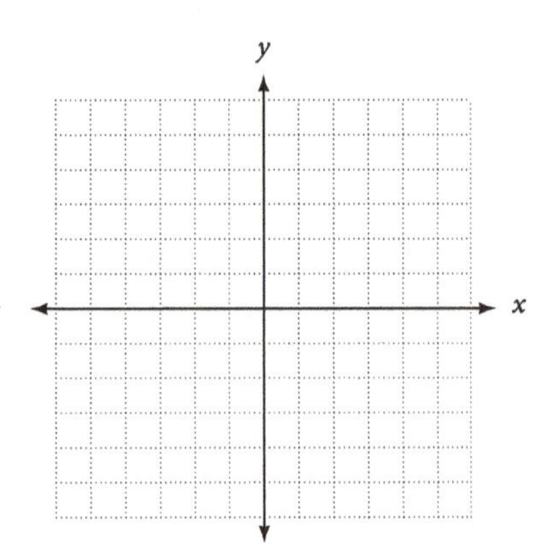

Matched Problems with Objectives

Second-Degree Inequalities and Nonlinear Systems

Section 10.3

Name ______________________

Date ______________________

Section ______________________

Directions Each problem below is similar to the example with the same number in your textbook. After reading through an example in your textbook, or watching one of the videos of that example on MathTV, try the matched problem to check your progress in this section. The shaded text is the learning objective associated with the matched problems that appear below the objective.

Objective A Graph second-degree inequalities. (Problem Set exercises 1 – 18 are similar.)

1. Graph: $x^2+y^2>9$

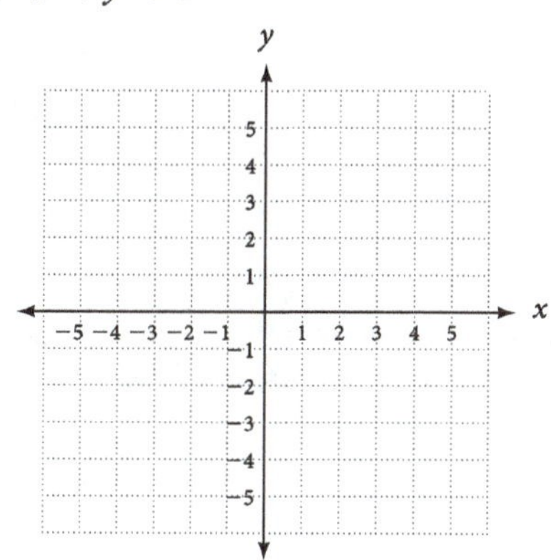

2. Graph: $y \geq x^2+3$

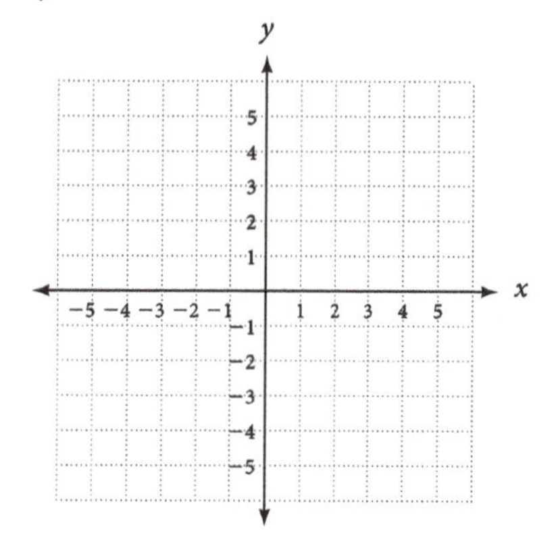

3. Graph: $9x^2-4y^2>36$

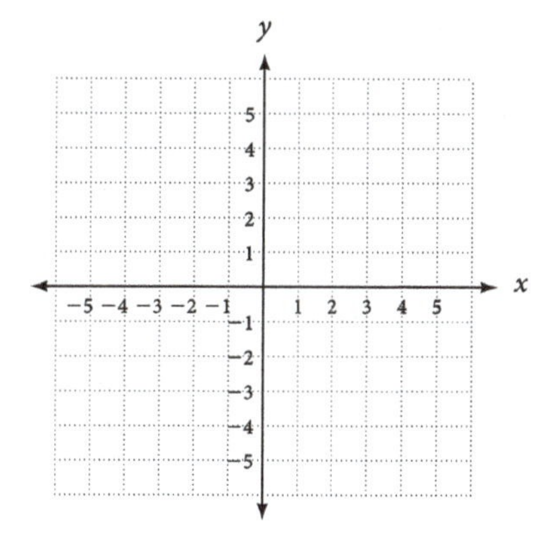

Matched Problems with Objectives

Second-Degree Inequalities and Nonlinear Systems

Section 10.3

Name ____________________

Date ____________________

Section ____________________

Objective B Solve systems of nonlinear equations. (Problem Set exercises 37 – 62 are similar.)

4. Solve the system:

$$x^2 + y^2 = 9$$
$$x - y = 3$$

5. Solve the system:

$$16x^2 - 4y^2 = 64$$
$$x^2 + y^2 = 4$$

6. Solve the system:

$$x^2 + y^2 = 4$$
$$y = x^2 - 4$$

7. One number is two less than the square of another number. The sum of the squares of the two numbers is 58. Find the two numbers.

SECTION 10.3

Matched Problems with Objectives

Second-Degree Inequalities and Nonlinear Systems

Section 10.3

Name ______________________

Date ______________________

Section ______________________

Objective C Graph the solution sets to systems of inequalities. (Problem Set exercises 19 –36are similar.)

8. Graph the solution set for the following system:

$x^2 + y^2 \geq 9$

$\frac{x^2}{4} + \frac{y^2}{25} \leq 1$

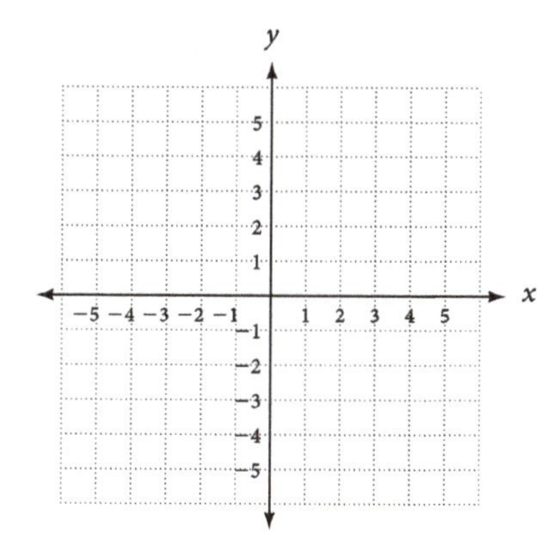

Answers

Chapter 1

Section 1.1

1. a. 23 **b.** 96 **c.** 125 **2. a.** 13, 16 **b.** $-2, -4$ **c.** $5, \frac{13}{2}$ **3. a.** 192, 768 **b.** $-56, 112$ **c.** $\frac{1}{81}, \frac{1}{243}$

4. 89 **5.** 54 **6.** 12 **7.** -7 **8.** 33 **9.**

-3.2 -0.5 $\frac{3}{4}$ 2.1

-3 -2 -1 0 1 $\sqrt{3}$ 2 3

10. a. 5 **b.** $-3, 5$ **c.** $-3, -2.1, \frac{1}{4}, 5$ **d.** $\sqrt{2}, \sqrt{7}$ **e.** $-3, -2.1, \frac{1}{4}, \sqrt{2}, \sqrt{7}, 5$

Section 1.2

1. $42x$ **2.** x **3.** $12x$ **4.** $4t$ **5.** 5 **6.** $15x+12$ **7.** $2x+1$ **8.** $0.05x+20$ **9.** $4x+3$
10. $-27x+45$ **11.** $-7a+14$ **12.** $11x$ **13.** $-24x^7$ **14.** $-432x^{17}$ **15.** $20x^4+40x^3+50x^2$
16. $6x^2-25x+14$ **17.** $12x^2+x-6$ **18.** $6x^3-13x^2y+8xy^2-3y^3$ **19.** $25x^2-20x+4$
20. $R=600p-100p^2$

Section 1.3

1. $9x+13$ **2.** $20y+40$ **3.** $13y+23$ **4.** $2x+17$ **5.** $x^2+4x-41$ **6.** $6x^2+3x+2$
7. $3x^2-9x+14$ **8. a.** 10 and 4 **b.** 9 and 9 **c.** 13 and 25 **9. a.** -2 **b.** -5 **c.** 0 **10. a.** -27
b. 38 **c.** $9-8y$

Section 1.4

1. $2^3 \cdot 3^2$ **2.** $\frac{3}{5}$ **3.** $(2x+y)(a+2b)$ **4.** $(3x+4)(2x+1)$ **5.** $(3x+7)(3x-7)$

6. $a^3+b^3=(a+b)(a^2-ab+b^2)$ and $a^3-b^3=(a-b)(a^2+ab+b^2)$ **7.** $(x+3)(x^2-3x+9)$

8. $3x^3(2x+3)(2x-3)$ **9.** $5x^2(x-3)^2$ **10.** $3(y^2+25)$ **11.** $(4a+1)(3a-4)$

12. $6x(x+1)(x^2-x+1)$ **13.** $9x^2y(x^2-2x+y^2)$ **14.** $2(4x+5)(a-3)$ **15.** $(5x+2)(x-2)(x+2)$

Section 1.5

1. 3 **2.** 9 **3.** 13 **4.** $\frac{1}{16}$ **5.** $-\frac{1}{27}$ **6.** $\frac{25}{16}$ **7. a.** 4 **b.** a^{10} **c.** $\frac{1}{x^4}$ **d.** $\frac{1}{m^2}$ **8.** a. 1

b. $3x^6y^5$ **9.** x^{10} **10.** $\frac{a^8}{4b^7}$ **11.** $\frac{8}{x^{14}y^2}$ **12.** $\frac{2}{3}, \frac{2}{5}$ **13. a.** 6 **b.** 6 **c.** 2 **d.** 6

Section 1.6

1. 0.439 miles per hour **2.** 648 miles per hour **3.** 46,523 deaths per week **4.** 8.4 cm/yr
5. 89.5 miles per hour **6.** 60 grams of calcium **7.** 152.5 cubic inches **8.** 5.67×10^4 **9.** 38,900
10. 1.5×10^5 or 150,000 **11.** 2.7×10^5 or 270,000 **12.** 2.4×10^8 or 240,000,000

Answers

Chapter 2

Section 2.1

1. 3 **2.** $-\frac{13}{24}$ **3.** 350 **4.** −30 **5.** 3 or 5 **6.** 0 or $\frac{1}{9}$ **7.** −6 or 7 **8.** $-\frac{1}{3}$, 0, or 3 **9.** 3, ±2 **10.** No solution **11.** All real numbers

Section 2.2

1. −4 **2.** $1.50 **3.** 2 miles per hour **4.** 1 second or 4 seconds **5.** $6 **6.** $x=\frac{P-2y-2z}{2}$ **7.** 34.54 ft/min **8.** $x=\frac{12}{a-2b}$ **9.** $y=\frac{c-ax}{b}$ **10.** $y=4x-17$ **11.** $r=\frac{A-P}{Pt}$

Section 2.3

1. The length is 22 inches and the width is 13 inches **2.** The price of the van was $18,000. **3.** One angle is 29° and the other is 61°. **4.** There is $4,500 invested at 3% and $5,500 invested at 5%. **5.** The sides are 6, 8, and 10. **6.** It will take 2.5 hours.

7.

Width	Length	Area (sq in)
1	$t=7-1=6$	6
2	$t=7-2=5$	10
3	$t=7-3=4$	12
4	$t=7-4=3$	12
5	$t=7-5=2$	10
6	$t=7-6=1$	6

Section 2.4

1. $(-8,\infty)$ **2.** $(-\infty,2]$ **3.** $(-\infty,3)$ **4.**

5.

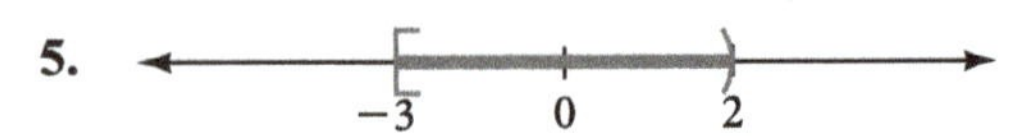

6.

$[-2,3]$ **7.** $x\le-3$ or $x\ge4$ **8.** $12 or less **9.** 15° to 25° C

Section 2.5

1. −6 or 6 **2.** −6 or 5 **3.** −10 or −25 **4.** No solution **5.** −1 or 1 **6.** $x=\frac{3}{2}$

Section 2.6

1. $-\frac{7}{3}<x<3$

2. $-9\le x\le 4$

3. $x<2$ or $x>12$

4. $x\le\frac{2}{5}$ or $x\ge2$

5. $-\frac{5}{2} < x < 1$

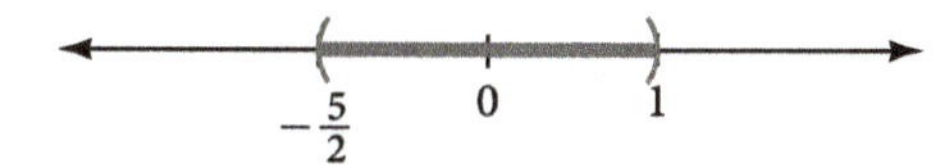

6. $x < -3$ or $x > 7$

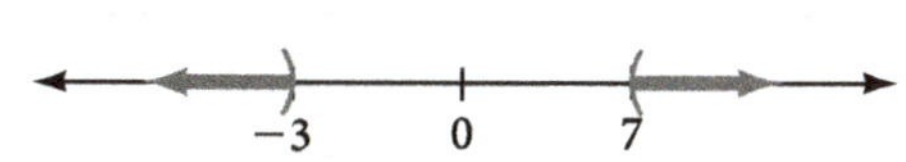

7. No solution

8. All real numbers

Answers

Chapter 3

Section 3.1

1.

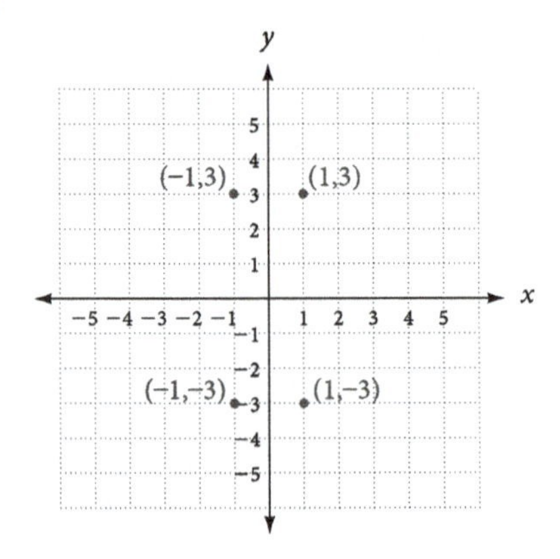

2.

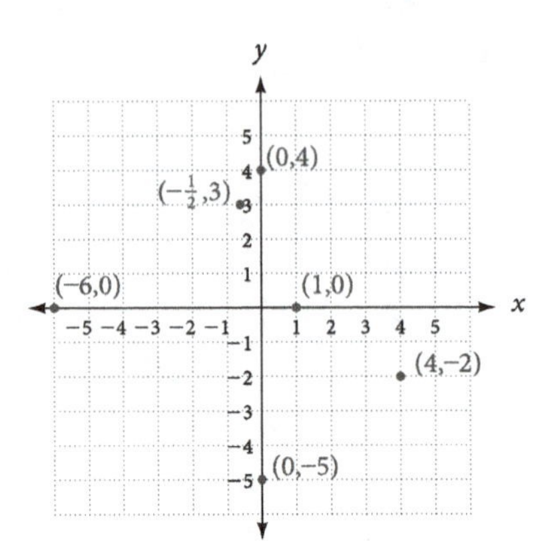

3.

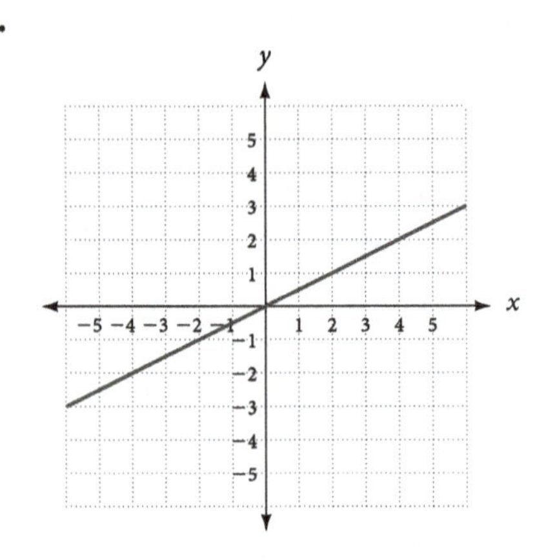

4.

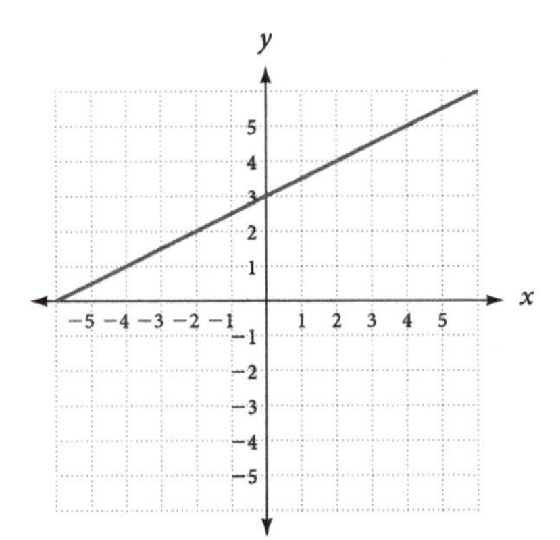

5.

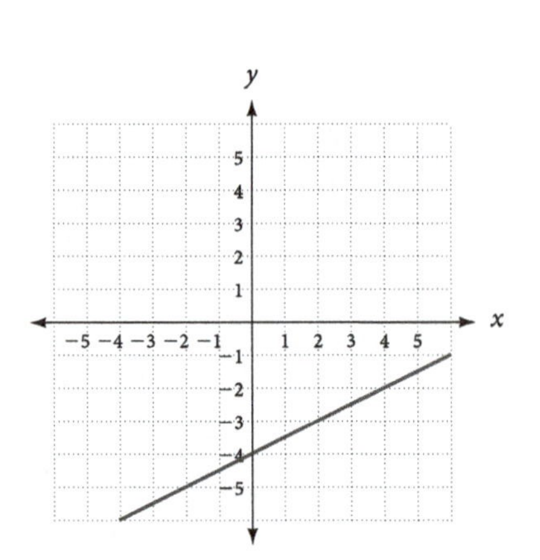

6. $y = 2x + 3$

7.

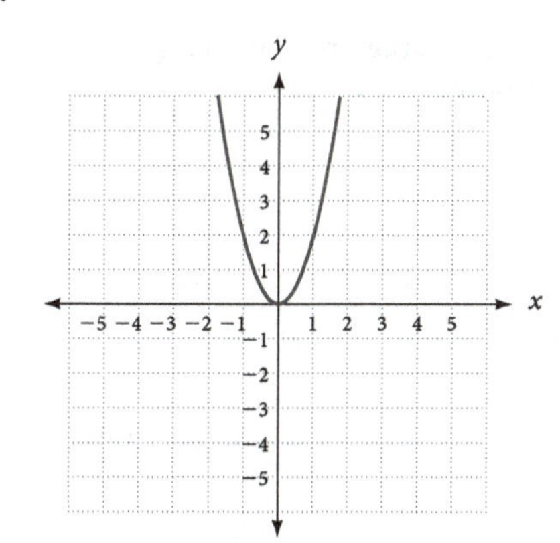

8.

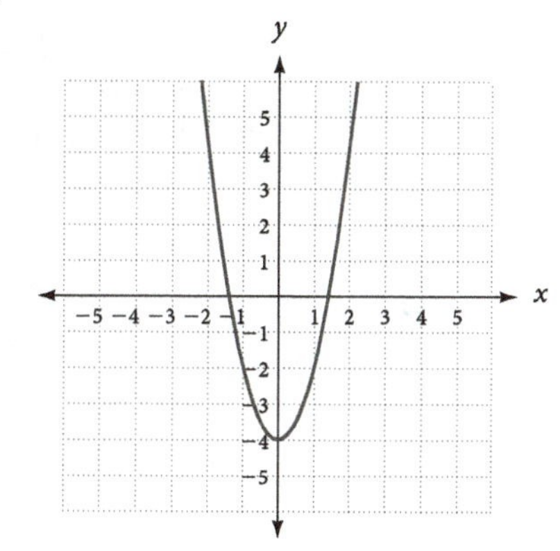

9.

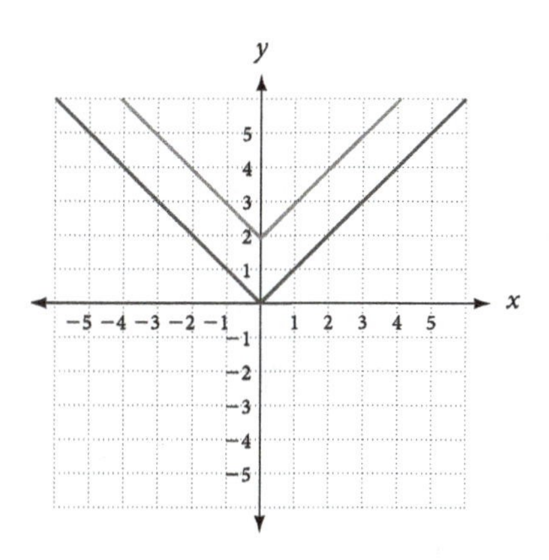

10. $(3,0)$ and $(0,-2)$

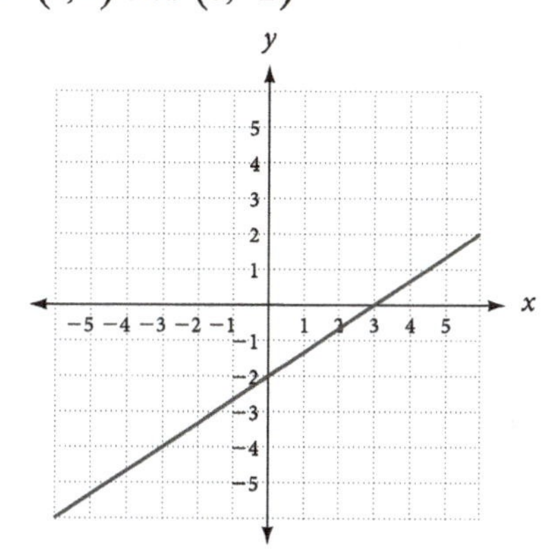

11. $(1,0), (-1,0)$, and $(0,-1)$

12.

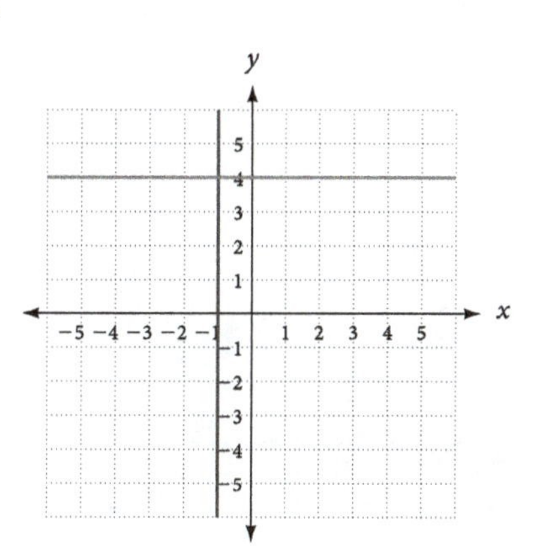

Section 3.2

1. 3

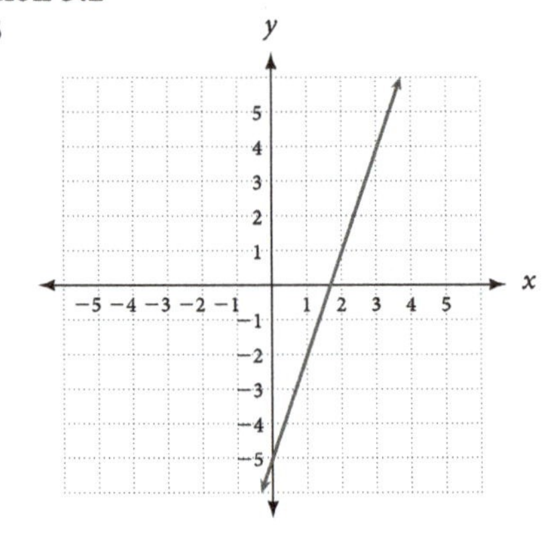

2. 3 **3.** 0 **4.** 8.4 cents per year which is the average rate of change of the price of a gallon of gasoline over a 20-year period of time.

5. 45 miles per hour

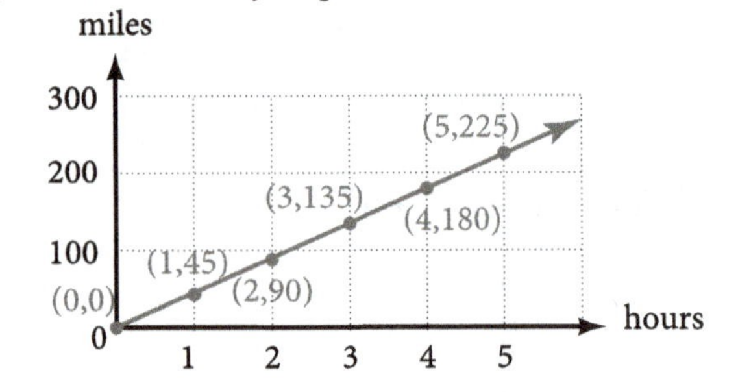

6. $-\frac{1}{2}$

Section 3.3

1. $y=\frac{2}{3}x+1$

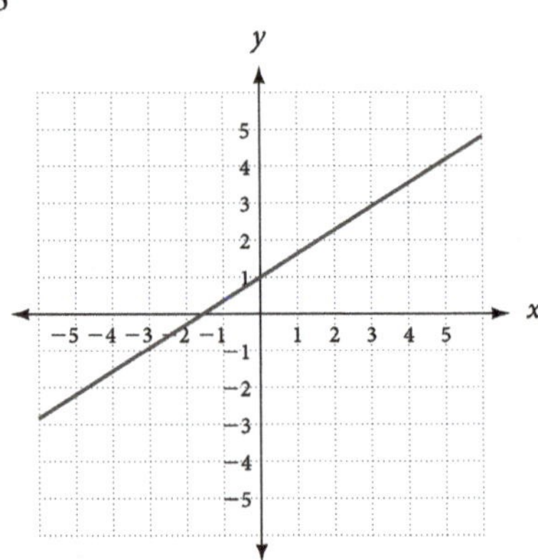

2. $m=\frac{4}{5}, \quad b=-\frac{7}{5}$

3.

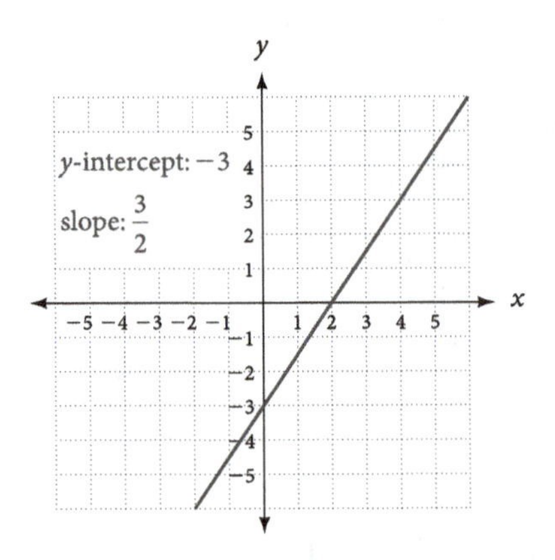

4. $y=3x+5$

5. $y=-2x+9$

6. $x+3y=9$

Section 3.4

1.

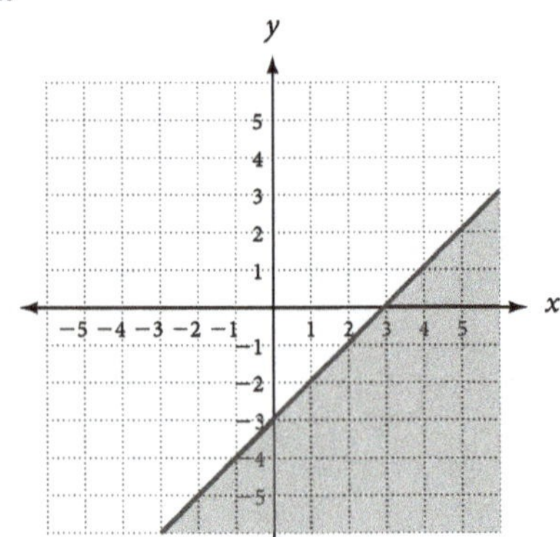

2.

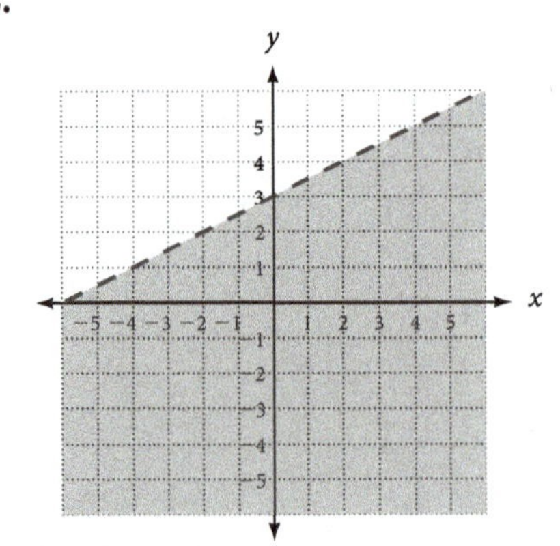

3.

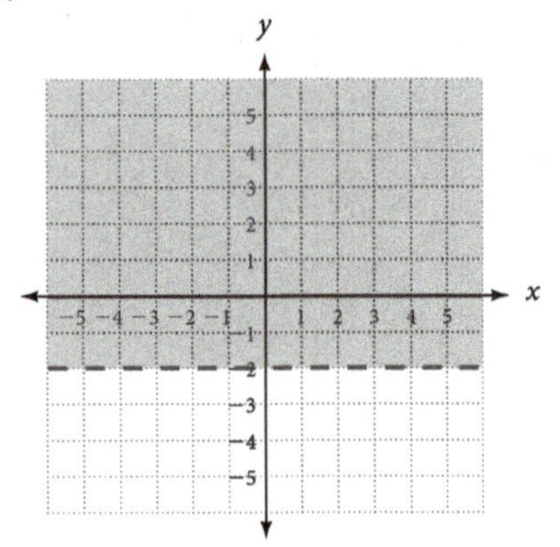

Section 3.5

1.

x	Function Rule $y=8x$	y
0	$y=8(0)$	0
10	$y=8(10)$	80
20	$y=8(20)$	160
30	$y=8(30)$	240
40	$y=8(40)$	320

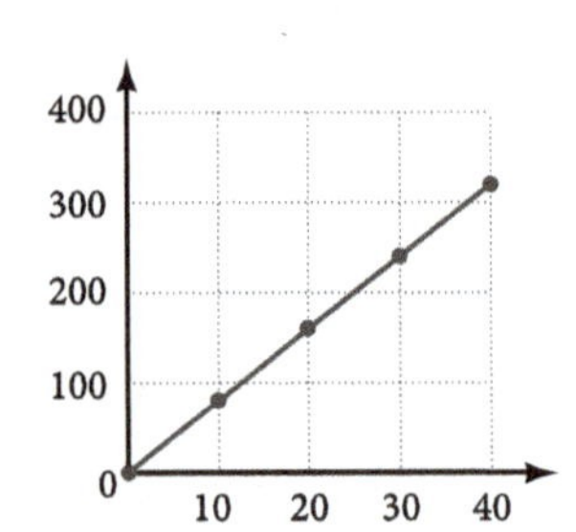

2. Domain: $\{0 \le x \le 40\}$

Range: $\{0 \le y \le 320\}$

3.

Time (sec) t	Function Rule $h = 48t - 16t^2$	Distance h
0	$h = 48(0) - 16(0)^2$	0
1	$h = 48(1) - 16(1)^2$	32
1.5	$h = 48(1.5) - 16(1.5)^2$	36
2	$h = 48(2) - 16(2)^2$	32
3	$h = 48(3) - 16(3)^2$	0

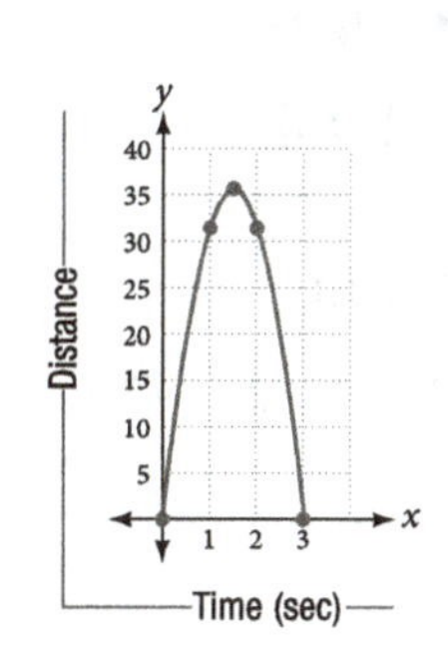

4. This relation is not a function.

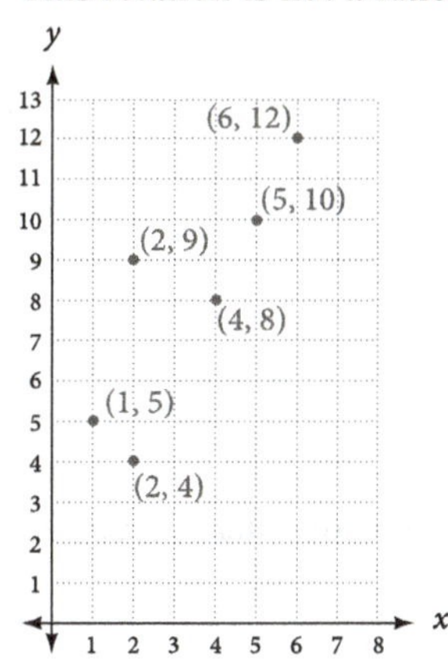

5. No, this is not the graph of a function.

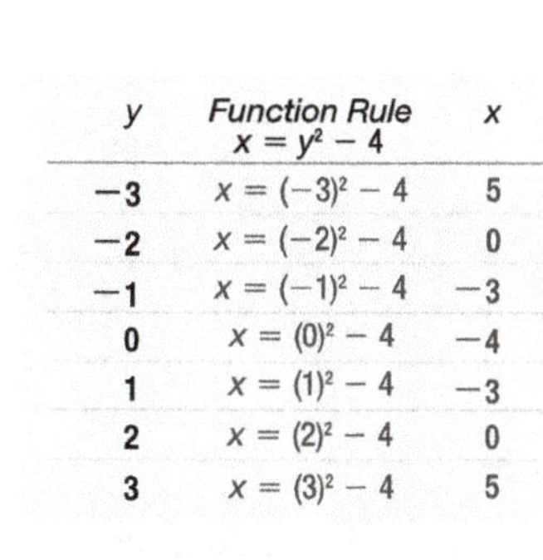

y	Function Rule $x = y^2 - 4$	x
−3	$x = (-3)^2 - 4$	5
−2	$x = (-2)^2 - 4$	0
−1	$x = (-1)^2 - 4$	−3
0	$x = (0)^2 - 4$	−4
1	$x = (1)^2 - 4$	−3
2	$x = (2)^2 - 4$	0
3	$x = (3)^2 - 4$	5

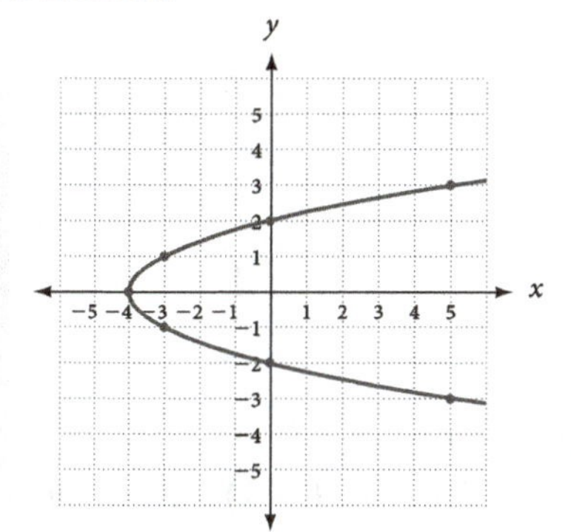

6. Yes, this is the graph of a function.

x	Function Rule $y = \lvert x\rvert - 3$	y
−3	$y = \lvert -3\rvert - 3$	0
−1	$y = \lvert -1\rvert - 3$	−2
0	$y = \lvert 0\rvert - 3$	−3
1	$y = \lvert 1\rvert - 3$	−2
3	$y = \lvert 3\rvert - 3$	0

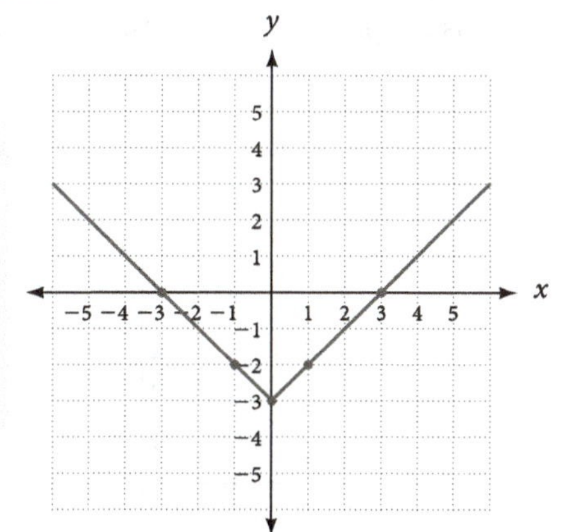

Section 3.6

1. a. 0 **b.** 40 **c.** 84 **2. a.** 11; it means that if Lorena runs a mile in 8 minutes, her average speed is 11 feet per second. **b.** 8; it means that if Lorena runs a mile in 11 minutes, her average speed is 8 feet per second. **3. a.** 40 ng/mL; after 5 days the concentration is 40 ng/mL. **b.** 20 ng/mL; after 10 days the concentration is 20 ng/mL. **4.** $C(5) = 10\pi \approx 31.4$ inches and $A(5) = 25\pi \approx 78.5$ square inches **5. a.** −3 **b.** 33 **c.** 13 **6. a.** 11 **b.** 22 **c.** −3 **d.** 1 **e.** $2a + 1$ **f.** $a^2 - 3$ **7. a.** 1 **b.** −3 **c.** 9 **8. a.** 243 **b.** 49

Section 3.7

1. 6 **2.** 100 feet **3.** 16 pound per square inch **4.** 32 **5.** 2.4 ohms

Section 3.8

1. a. $x^2 + x - 2$ **b.** $x^2 - x - 6$ **c.** $x^3 + 2x^2 - 4x - 8$ **d.** $x - 2$ **2. a.** $3x^2 - 7x - 6$ **b.** $3x^2 - 10x - 8$ **c.** $9x^3 - 24x^2 - 44x - 16$ **d.** $x - 4$ **3. a.** 27 **b.** 14 **c.** −9 **d.** 9 **4. a.** $x^2 + 3x - 4$ **b.** $x^2 - 5x + 4$

Answers

Chapter 4

Section 4.1

1. $(2,-3)$ **2.** $(-2,1)$ **3.** $\left(\frac{13}{22},-\frac{1}{22}\right)$ **4.** $\varnothing$ **5.** Lines coincide. **6.** $(6,4)$ **7.** $(2,5)$ **8.** $(1,3)$

9.

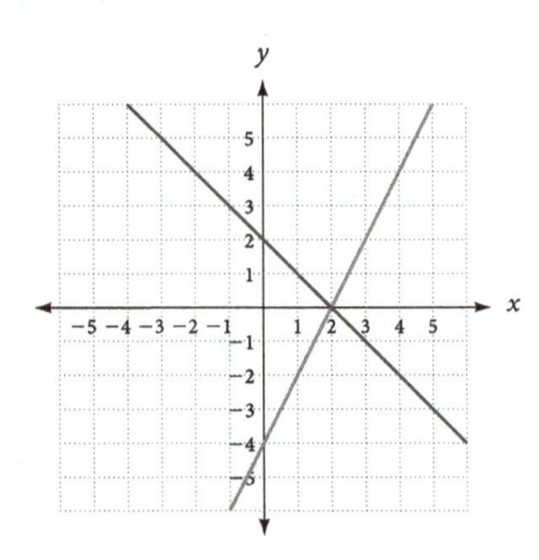

Section 4.2

1. $(1,2,-3)$ **2.** $(1,1,1)$ **3.** No unique solution; it is a dependent system. **4.** No solution: it is an inconsistent system.
5. $(4,-2,3)$

Section 4.3

1. 5 and 9 **2.** 340 adult tickets and 410 children's tickets **3.** \$5,000 at 6% and \$7,000 at 7% **4.** 4 gallons of 30% and 12 gallons of 70% **5.** Speed of boat is 8 mph and current is 2 mph **6.** 11 nickels, 3 dimes, and 1 quarter
7. $C=\frac{5}{9}(F-32)$ or $C=\frac{5}{9}F-\frac{160}{9}$

Section 4.4

1.

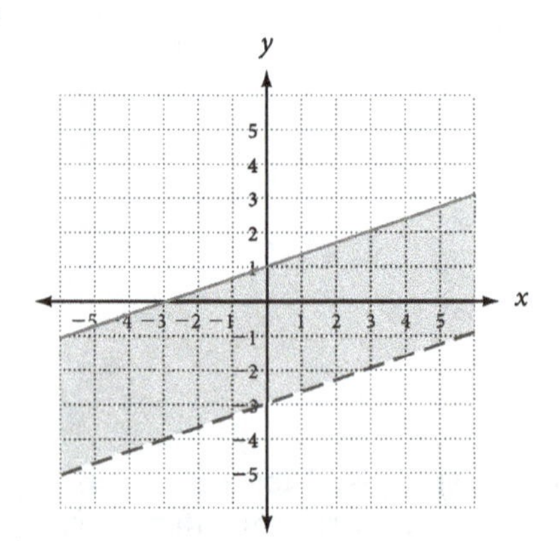

2.

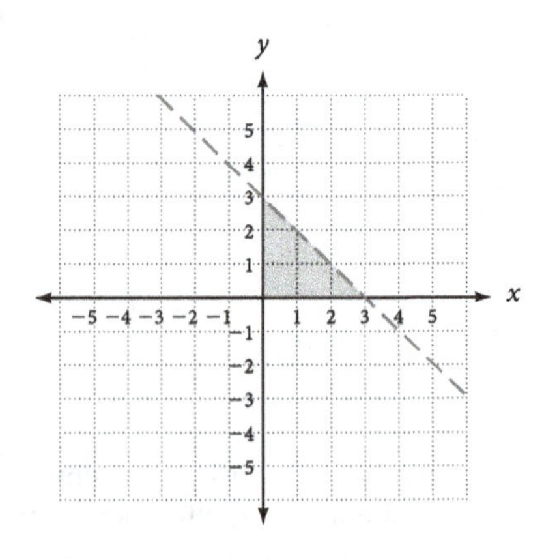

3.

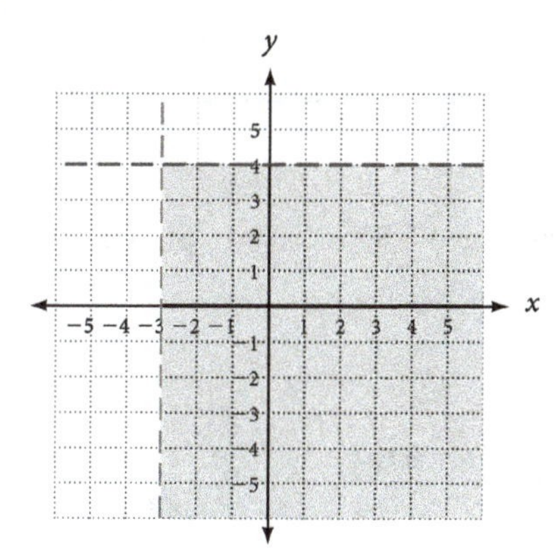

4.

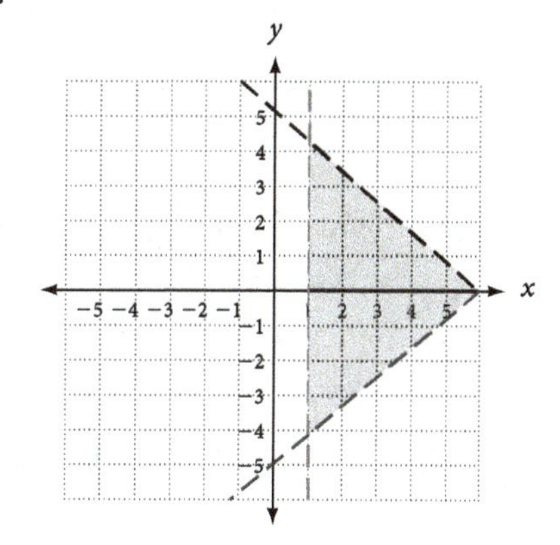

5. The horizontal line would move down to $y = 300$ from $y = 500$.

Answers

Chapter 5

Section 5.1

1. $x-3$ **2.** $\frac{y-3}{y-2}$ **3.** $\frac{a+1}{2}$ **4.** 1 **5.** -1 **6.** $\frac{-1}{x+7}$ **7. a.** -2 **b.** 0 **c.** 4 **d.** Undefined **8. a.** $x \neq 1$ **b.** $x \neq -2$ **c.** $x \neq \pm 1$ **9.** 2 **10.** $x+a$ **11.** 3

Section 5.2

1. $\frac{1}{3}$ **2.** $6x^2y^4$ **3.** $\frac{1}{(x-5)^2}$ **4.** $\frac{y(y-1)}{y+1}$ **5.** $\frac{3}{2}$ **6.** $6x^2y^2$ **7.** $\frac{y^2}{x^2-xy+y^2}$ **8.** $\frac{(a+4)(a+3)}{a-4}$ **9.** $\frac{x-y}{x+y}$ **10.** $3(x+3)$

Section 5.3

1. $\frac{1}{2}$ **2.** $\frac{1}{x-3}$ **3.** 1 **4.** $\frac{59}{105}$ **5.** $\frac{1}{(x+4)(x+2)}$ **6.** $\frac{x+2}{2(x+3)}$ **7.** $\frac{x-1}{(x+1)(x+2)}$ **8.** $x+3$ **9.** $\frac{10x+23}{5x-1}$ **10.** $\frac{1}{x}+\frac{1}{3x}=\frac{4}{3x}$

Section 5.4

1. $\frac{4}{5}$ **2.** $\frac{3-x}{3+x}$ **3.** $\frac{1}{(x+4)(x-5)}$ **4.** $\frac{x+3}{x+2}$ **5.** $\frac{10x+23}{5x-1}$

Section 5.5

1. $-\frac{3}{2}$ **2.** 1 **3.** No solution **4.** 9 **5.** $-2, 4$ **6.** 7 **7.** $y=\frac{x+2}{x-1}$ **8.** $x=\frac{ab}{b-a}$

9.

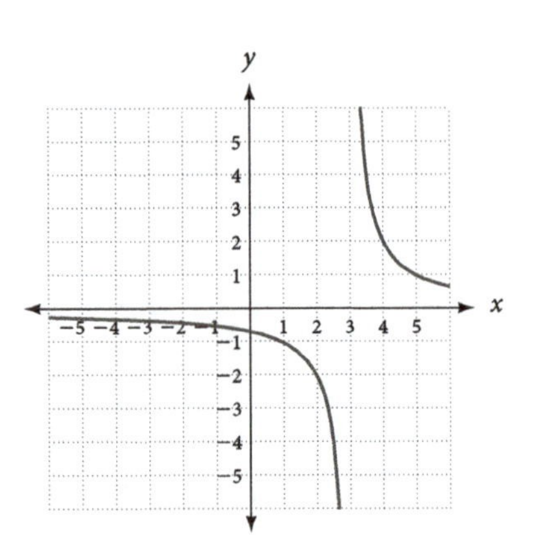

10.

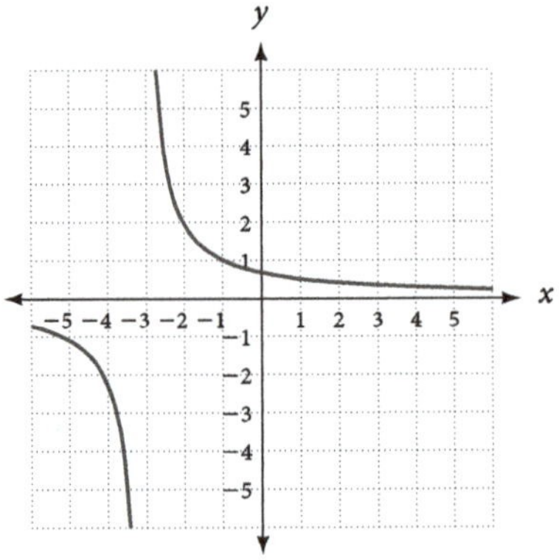

Section 5.6

1. 1 and 3 **2.** 5 mph **3.** 6 mph **4.** 60 hours **5.**

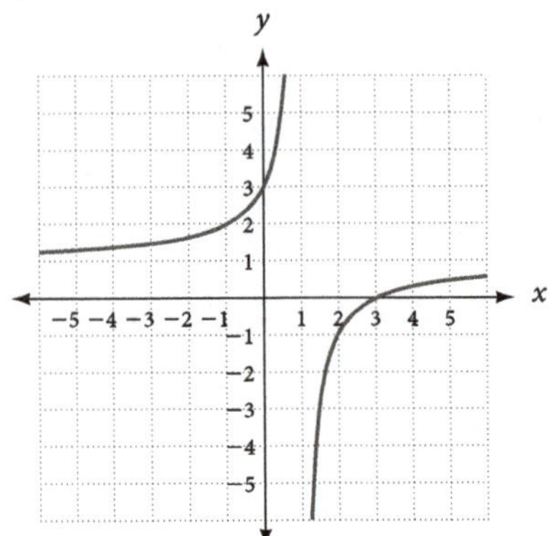

Section 5.7

1. $2x^3-3x^2+4x$ **2.** $-3xy^5+9x^2y$ **3.** $\frac{3a}{2}+1+\frac{2}{a}+\frac{1}{2a^2}$

4. $2x-3y$ **5.** $215+\frac{3}{5}$ **6.** $3x+1+\frac{2}{x-3}$

7. $3x^2+6x+15+\frac{31}{x-2}$ **8.** $x-7y$ **9.** $(x-2)(x-3)(x+1)$

Answers

Chapter 6

Section 6.1

1. $-6, 6$ **2.** -4 **3.** Not a real number **4.** -2 **5.** -1 **6.** Not a real number **7.** $9a^2b^4$ **8.** $2xy^3$ **9.** $3ab^2$

10. 3 **11.** 3 **12.** -7 **13.** Not a real number **14.** $\frac{4}{5}$ **15.** $2xy^3$ **16.** $3ab^2$ **17.** 27 **18.** 8 **19.** $\frac{1}{4}$ **20.** $\frac{27}{8}$

21. $x^{\frac{3}{4}}$ **22.** $y^{\frac{1}{2}}$ **23.** $z^{\frac{1}{12}}$ **24.** $\frac{1}{a^4b^2}$ **25.** $\frac{1}{x^2y^{28}}$

Section 6.2

1. $3\sqrt{2}$ **2.** $5xy\sqrt{2y}$ **3.** $3ab\sqrt[3]{2a}$ **4.** $5x^2y^4\sqrt{3x}$ **5.** $2a^2bc\sqrt[4]{3b}$ **6.** $\frac{\sqrt{5}}{3}$ **7.** $\frac{\sqrt{6}}{3}$ **8.** $\frac{5\sqrt{2}}{2}$ **9.** $\frac{3\sqrt{10xy}}{2y}$

10. $\frac{5\sqrt[3]{3}}{3}$ **11.** $\frac{4xy^2\sqrt{21xz}}{7z}$ **12.** $4|x|$ **13.** $5|x|\sqrt{x}$ **14.** $|x+5|$ **15.** $|x|\sqrt{2x+7}$ **16.** -3 **17.** -1

Section 6.3

1. $5\sqrt{5}$ **2.** $26\sqrt{2}$ **3.** $-3x\sqrt{2y}$ **4.** $21b\sqrt[3]{a^2b}$ **5.** $\frac{8\sqrt{5}}{15}$

6. First we construct a golden rectangle from a square of side 6.

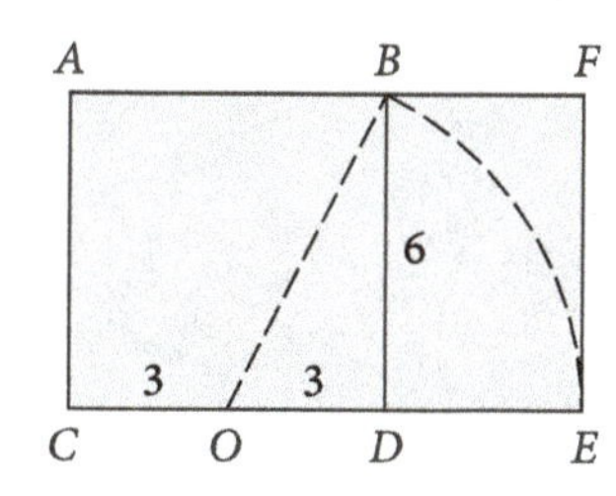

The length of the diagonal OB is found from the Pythagorean theorem to be $3\sqrt{5}$. The ratio of the length to the width for the rectangle is the golden ratio, $\frac{1+\sqrt{5}}{2}$.

Section 6.4

1. $35\sqrt{33}$ **2.** $3\sqrt{10}-8$ **3.** $-19-2\sqrt{14}$ **4.** $x+10\sqrt{x}+25$ **5.** $25a-30\sqrt{ab}+9b$ **6.** $x+4-2\sqrt{x+3}$ **7.** 2

8. $\frac{3\sqrt{7}+3\sqrt{3}}{4}$ **9.** $19-6\sqrt{10}$ **10.** First we construct a golden rectangle from a square of side 4. Next, we find expressions for the length and width of the smaller rectangle.

$$\text{Length} = EF - 4$$

$$\text{Width} = DE - 2\sqrt{5} - 2$$

Next, we find the ratio of length to width.

$$\text{Ratio of length to width} = \frac{EF}{DE} = \frac{4}{2\sqrt{5}-2} = \frac{2}{\sqrt{5}-1}$$

Section 6.5

1. 6 **2.** No solution **3.** 5 **4.** 8 **5.** 9 **6.** 0 **7.** – 2 **8.** 5

9. $y = \sqrt{x} + 3$ $y = \sqrt{x+3}$

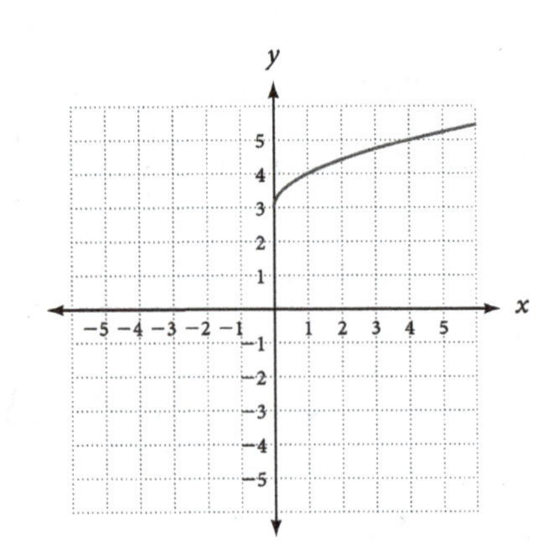

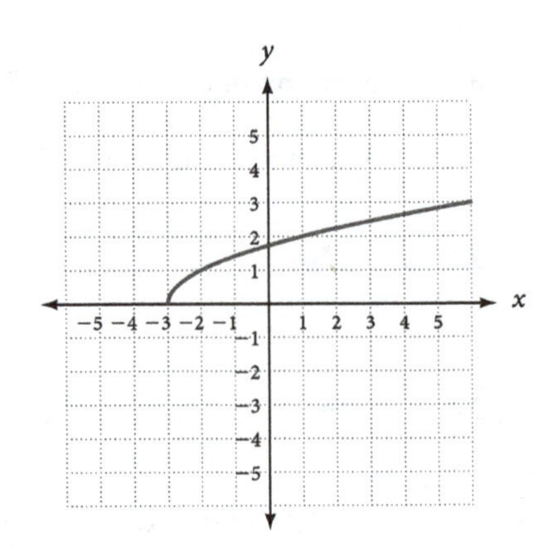

Section 6.6

1. $6i$ **2.** $-8i$ **3.** $3i\sqrt{2}$ **4.** $-i\sqrt{19}$ **5.** 1 **6.** $-i$ **7.** -1 **8.** $x = 2$ and $y = -\frac{1}{2}$ **9.** $x = 3$ and $y = 2$ **10.** $5 + 2i$

11. $2 + 2i$ **12.** $-1 + i$ **13.** $14 - 5i$ **14.** $9 - 6i$ **15.** $-12 + 16i$ **16.** 34 **17.** $-\frac{4}{29} + \frac{19}{29}i$ **18.** $2 - 3i$

Answers

Chapter 7

Section 7.1

1. $\frac{2}{3}, -2$ **2.** $\frac{3 \pm 5i\sqrt{2}}{4}$ **3.** $-5 \pm 2\sqrt{5}$ **4.** $-4, 1$ **5.** $\frac{3 \pm i\sqrt{31}}{10}$ **6.** 4 inches and $+2\sqrt{3}$ inches **7.** 4,222 feet

Section 7.2

1. $-\frac{1}{2}, -\frac{2}{3}$ **2.** $\frac{-3 \pm \sqrt{15}}{3}$ **3.** $-2 \pm 2i$ **4.** $\frac{-3 \pm 3\sqrt{3}}{4}, \frac{3}{2}$ **5.** $\frac{1}{2}$ second and $\frac{3}{2}$ seconds **6.** 130 or 140 copies of the program

Section 7.3

1. Two rational solutions **2.** One rational solution **3.** Two non-real complex solutions **4.** Two irrational solutions **5.** $k = \pm 12$ **6.** $t^3 - 3t^2 - 4t + 12 = 0$ **7.** $20x^2 + 11x - 3 = 0$

Section 7.4

1. $0, 7$ **2.** $\pm\frac{i\sqrt{3}}{3}, \frac{\sqrt{10}}{2}$ **3.** 16 **4.** $t = \frac{5 \pm \sqrt{25 + 16h}}{16}$

Section 7.5

1.

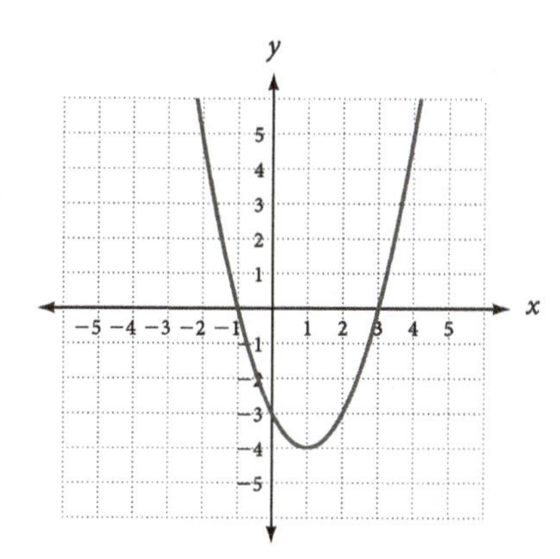

2.

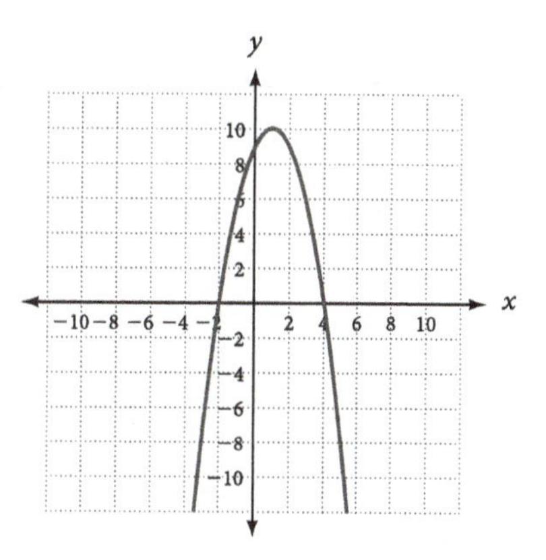

3.

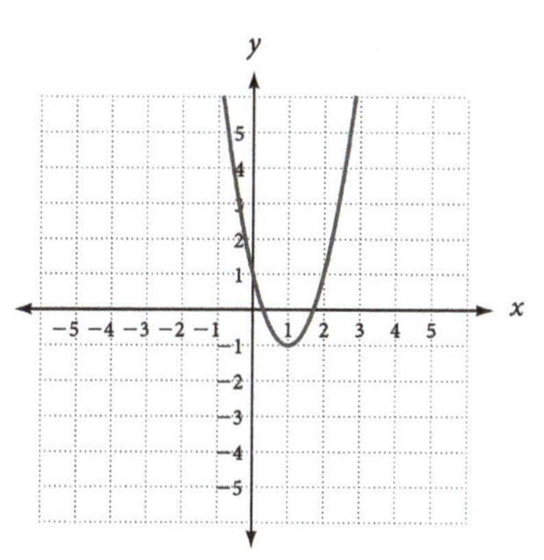

4.

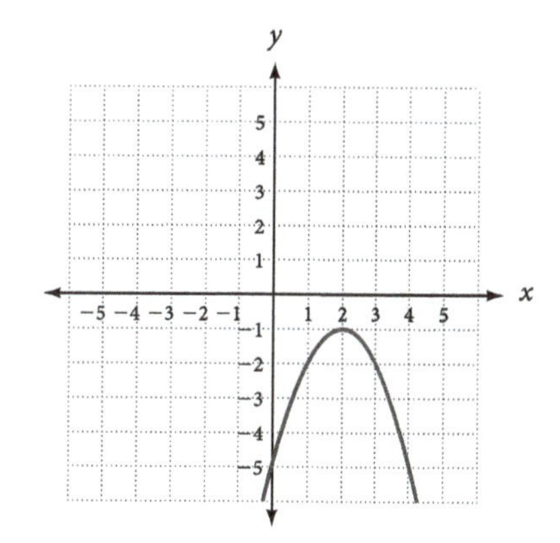

5. $y = 3{,}200$

6. $R = -200p^2 + 800p$ Maximum revenue is $800 when the price is $2 each.

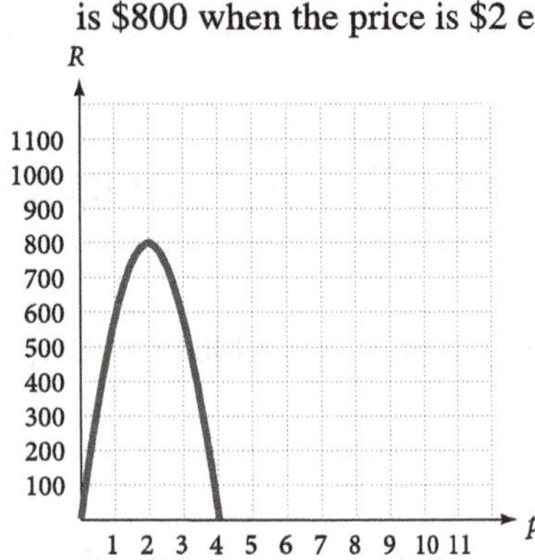

7. $y = -\frac{1}{80}(x - 80)^2 + 80$

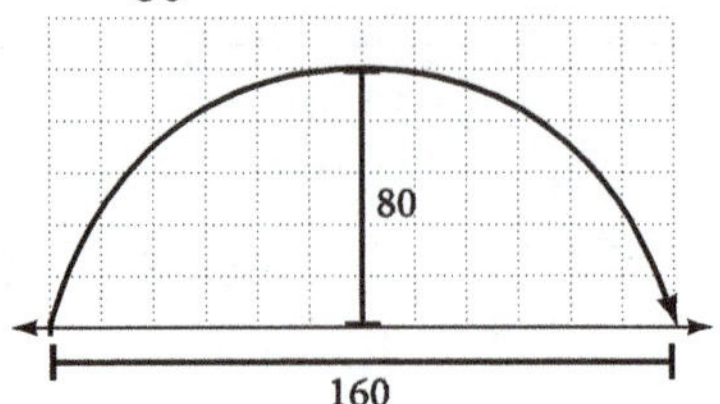

Section 7.6

1. $-5 \le x \le 3$

−5 0 3

2. $x < -2$ or $x > \frac{1}{2}$

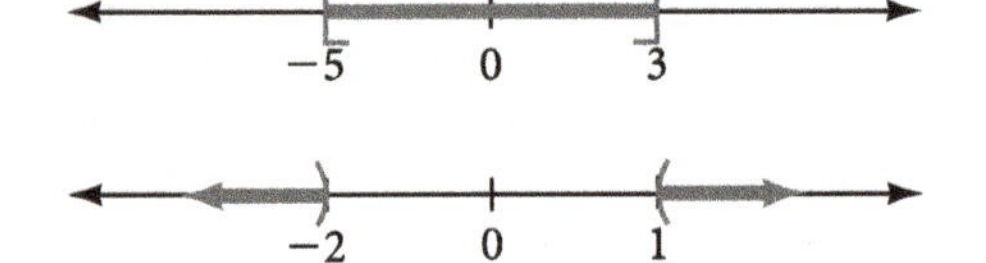

3. No solution

4. $-3 \le x < 1$

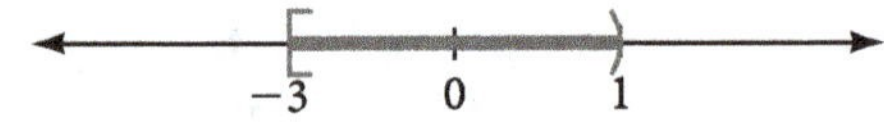

5. $-1 < x < 4$ or $x > 14$

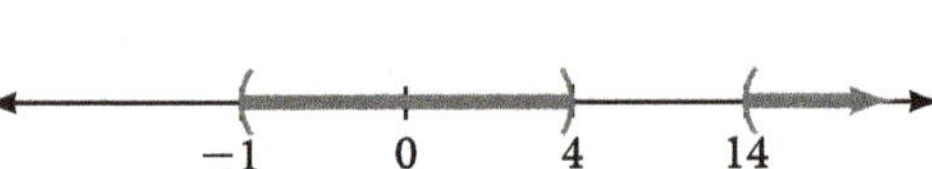

Answers

Chapter 8

Section 8.1

1. **a.** 1 **b** 4 **c.** 16 **d.** 64 **e.** $\frac{1}{4}$ **f.** $\frac{1}{16}$

2. $A(12) \approx 424.3$, $A(24) = 150$

3.

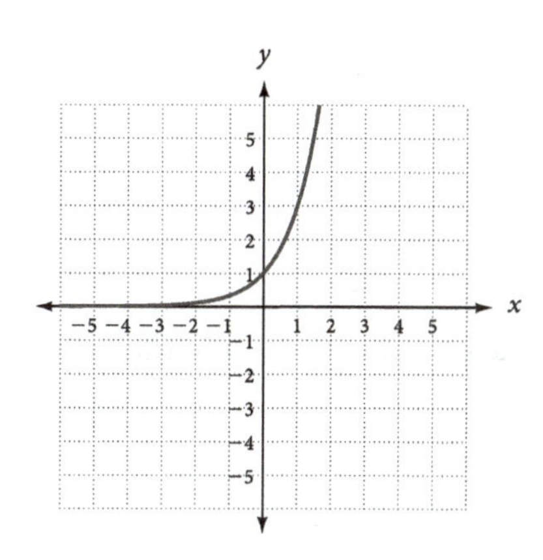

4.

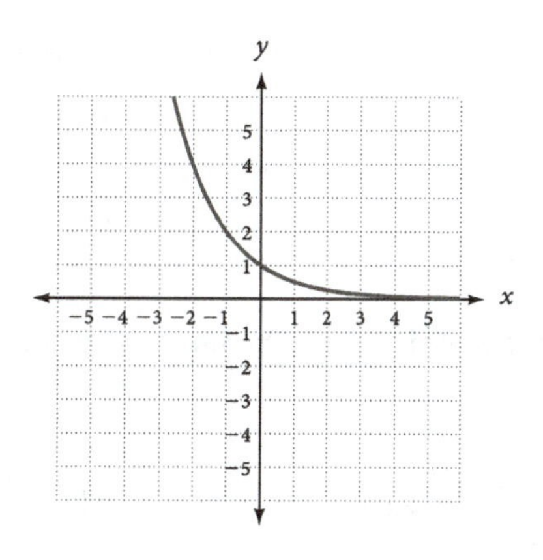

5. **a.** $A = 600\left(1+\frac{.06}{4}\right)^{4t}$

   b. \$808.11

   c. About 8.5 years

6. \$809.92

Section 8.2

1. $f^{-1}(x)=\frac{x-1}{4}$

2. $f^{-1}(x)=\pm\sqrt{x-1}$

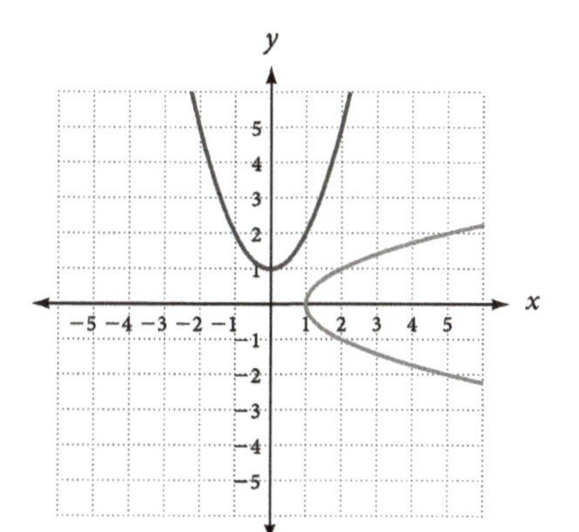

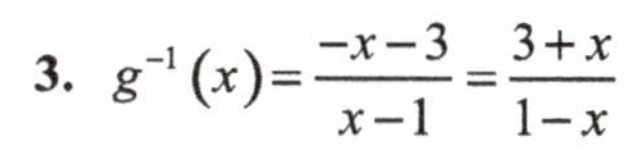

3. $g^{-1}(x)=\frac{-x-3}{x-1}=\frac{3+x}{1-x}$

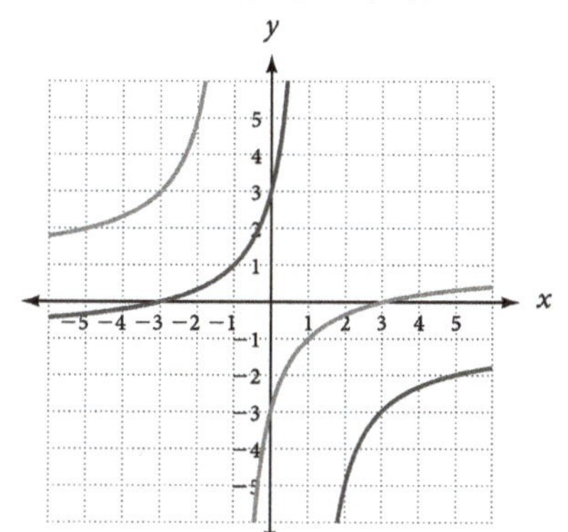

4.

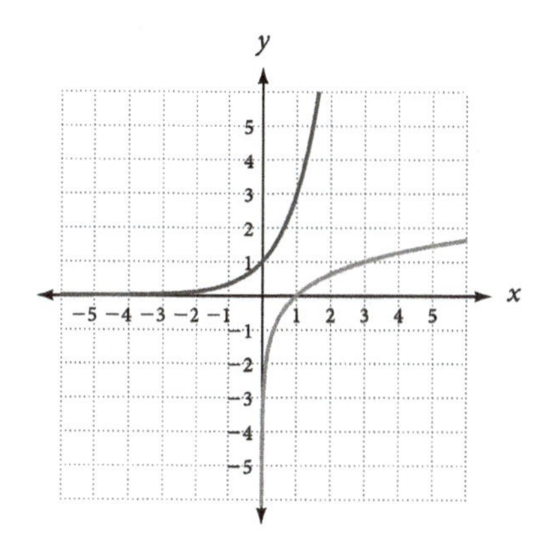

Section 8.3

1. $2^3 = x, x = 8$ **2.** $\sqrt{5}$ **3.** $\frac{3}{2}$ **4.**

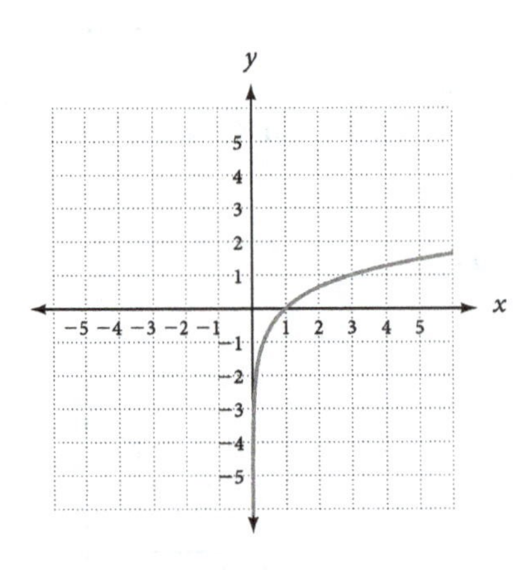

5. a 3 **b.** 3 **c.** 1 **d.** 0 **e.** 0

6. We can say that the shock wave is 1,000,000 times greater than the smallest shock wave measurable on a seismograph.

Section 8.4

1. $\log_3 5 + \log_3 a - \log_3 b$ **2.** $2\log_{10} x - \frac{1}{3}\log_{10} y$ **3.** $\log_4 \frac{x^3 y}{z^2}$ **4.** 1

Section 8.5

1. 4.4409 **2.** −2.4078 **3.** −2.0214 **4.** 9519.19 **5.** 0.0257 **6.** 3.16×10^5 **7.** 7.9×10^{-5} **8.** 4.7 **9. a.** 2 **b.** 4 **c.** −2 **d.** x **10.** $\ln P + rt$ **11. a.** 3.5553 **b.** −1.6094 **c.** 3.8918

Section 8.6

1. −0.7945 or −0.7944, depending on whether you round your intermediate answers **2.** 6.64 years **3.** 1.4728 or 1.4729, depending on whether you round your intermediate answers **4.** Approximately 17.04 years later

Answers

Chapter 9

Section 9.1

1. 1, 4, 7, 10 **2.** $\frac{3}{3}=1, \frac{3}{4}, \frac{3}{5}, \frac{3}{6}=\frac{1}{2}$ **3.** $a_5=\frac{1}{50}, a_6=-\frac{1}{72}$ **4.** 3, 10, 31, 94 **5.** $a_n=3n$ **6.** $a_n=\frac{n+2}{n+3}$

Section 9.2

1. $6+10+14+18+22=70$ **2.** $\frac{1}{4}-\frac{1}{8}+\frac{1}{16}-\frac{1}{32}=\frac{5}{32}$ 3. $(x^2+2)+(x^3+2)+(x^4+2)+(x^5+2)$ 4. $\sum_{i=1}^{5}(3i)$

5. $\sum_{i=1}^{4}(2i^2)$ **6.** $\sum_{i=1}^{4}\left(\frac{x+2i}{x^{2i}}\right)$

Section 9.3

1. $d=3$ **2.** $d=-4$ **3.** $d=-\frac{1}{2}$ **4.** $a_n=6n-1$ **5.** $a_n=3n-2$ **6.** $S_{10}=215$

Section 9.4

1. $r=3$ **2.** $r=\sqrt{2}$ **3.** $a_n=-6(-2)^{n-1}$ or $a_n=3(-2)^n$ **4.** $a_{10}=\frac{1}{256}$ **5.** $a_n=5(3)^{n-1}$ **6.** $S_{10}=7,324,218$

7. 4 **8.** $a_1=0.6$ and $r=0.1$, $S_\infty=\frac{0.6}{1-0.1}=\frac{0.6}{0.9}=\frac{2}{3}$

Answers

Chapter 10

Section 10.1

1. $2\sqrt{13}$ **2.** 2, 4 **3.** $(x-4)^2+(y+3)^2=4$ **4.** $x^2+y^2=25$

5. (3,4) $r=3$

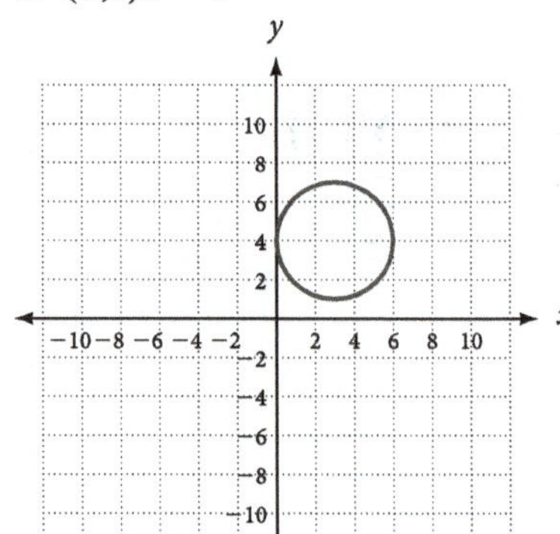

6.

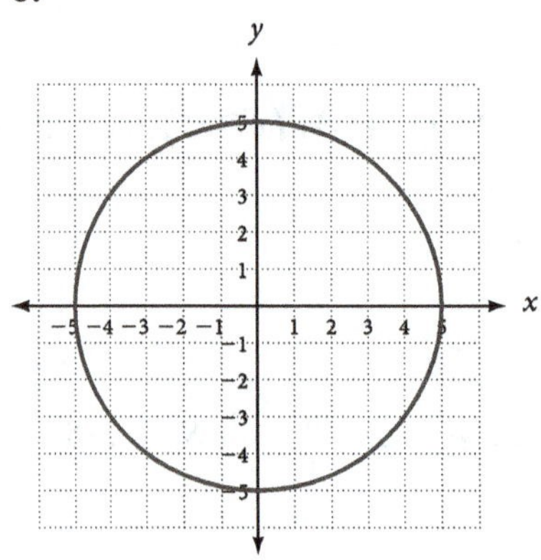

7.

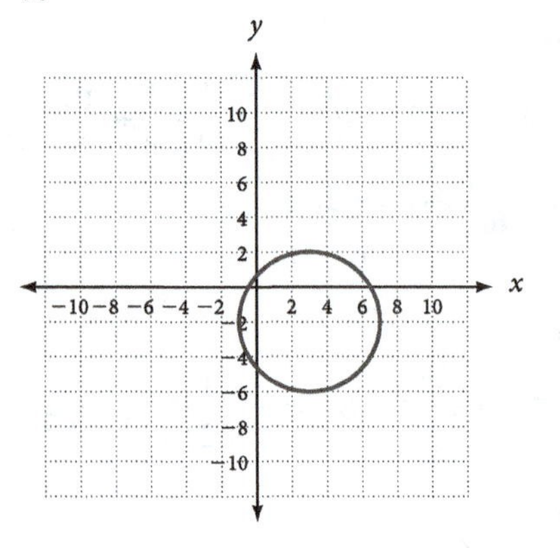

Section 10.2

1.

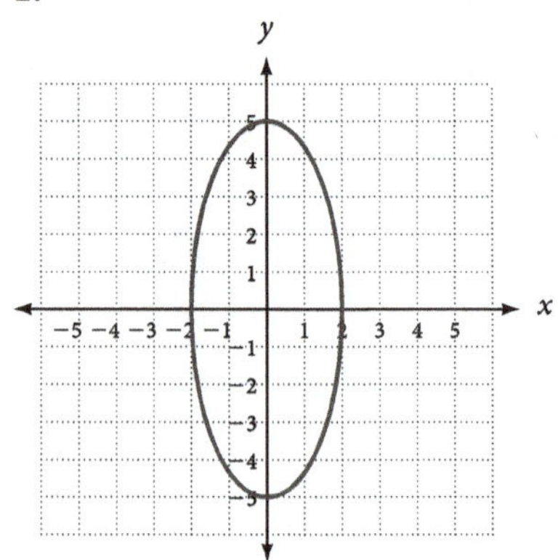

2.

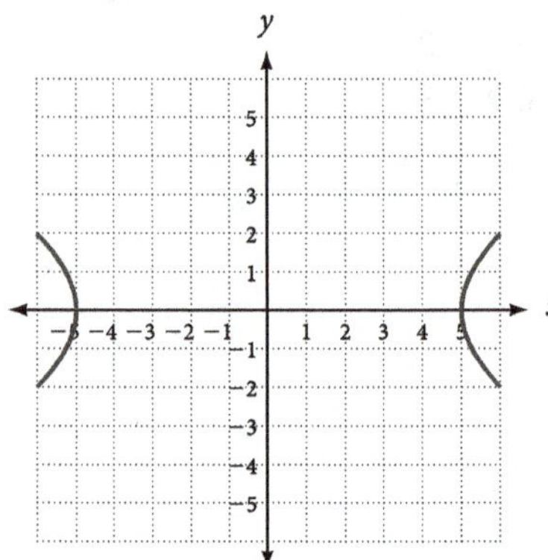

3.

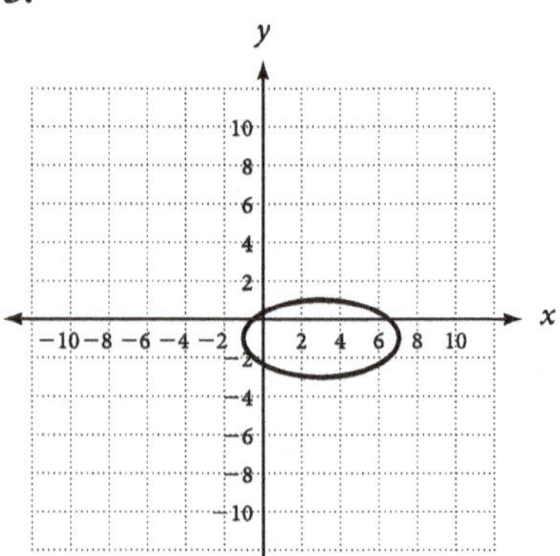

4.

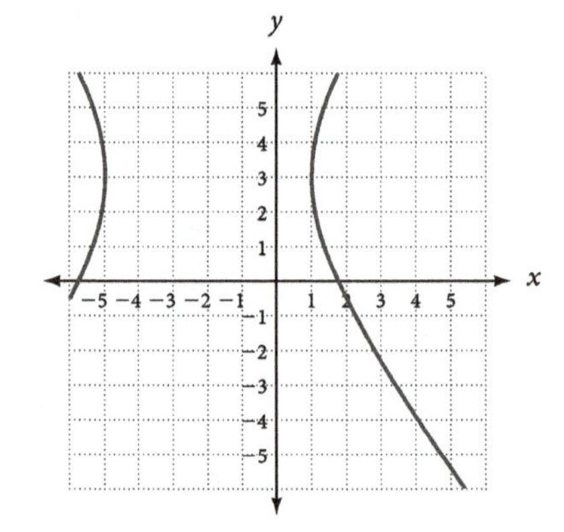

Section 10.3

1.

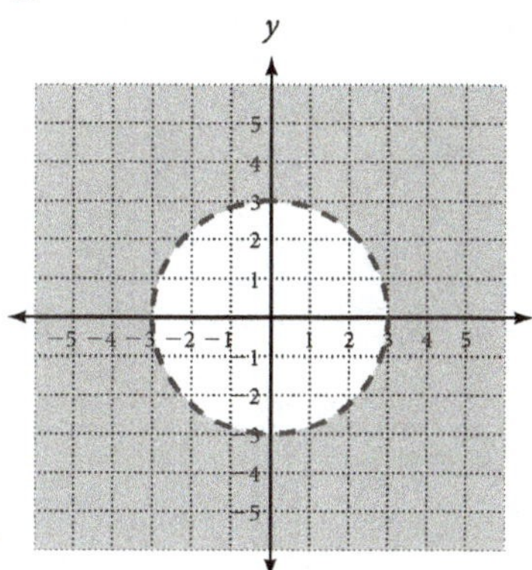

2.

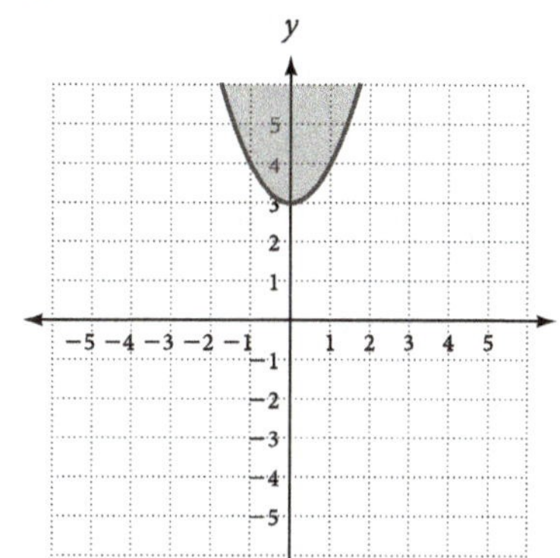

3.

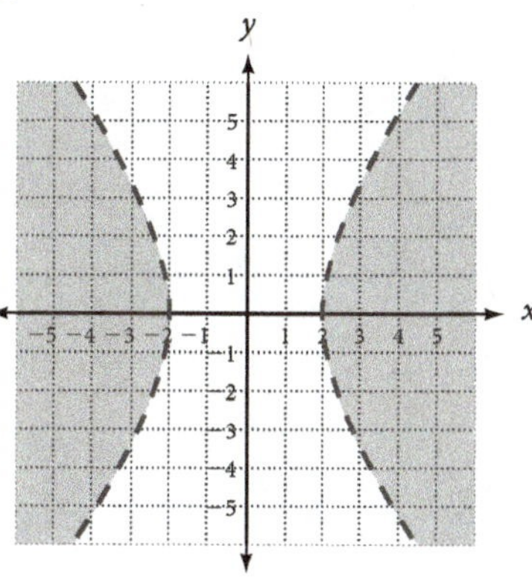

4. $(3,0),(0,-3)$ **5.** $(2,0),(-2,0)$ **6.** $(2,0),(-2,0),(\sqrt{3},-1),(-\sqrt{3},-1)$ **7.** 3, 7 and −3, 7

8.

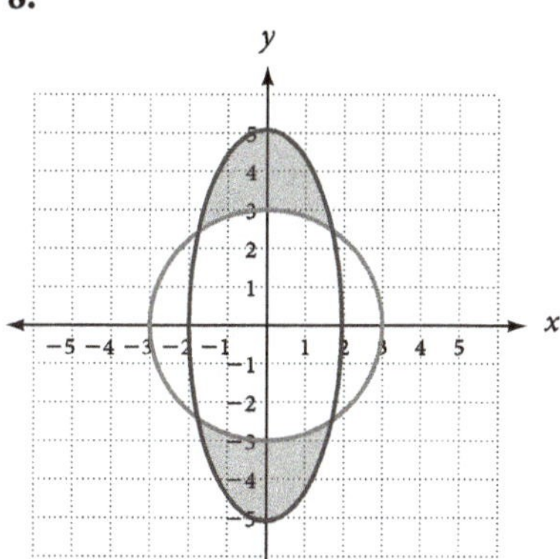

*xyz*textbooks

XYZ Textbooks was founded by Charles "Pat" McKeague in order to provide affordable textbooks to college students. Current textbook prices can be upwards of $150 per book. We offer all our books at a reasonable price, with professional, accessible service. Study skills, success skills, and common mistakes are integrated with all our material.

visit xyztextbooks.com for more information

*xyz*homework

With over 5,000 developmental math exercises correlated section by section to our textbooks, XYZ Homework provides virtually unlimited practice. From these questions, ready-to-use assignments providing instant feedback have been created to get you started in just a few clicks.

visit xyzhomework.com for more information

Current Titles

Essentials of Basic Mathematics

1. Whole Numbers
2. Fractions 1: Multiplication and Division
3. Fractions 2: Addition and Subtraction
4. Decimals
5. Ratio and Proportion
6. Percent

Introductory Mathematics

1. Whole Numbers
2. Fractions 1: Multiplication and Division
3. Fractions 2: Addition and Subtraction
4. Decimals
5. Ratio and Proportion
6. Percent
7. Measurement
8. Geometry
9. Introduction to Algebra
10. Solving Equations

Basic Mathematics with Early Integers

R. Whole Numbers
1. Introduction to Algebra
1. Fractions 1: Multiplication and Division
2. Fractions 2: Addition and Subtraction
3. Decimals
4. Ratio and Proportion
5. Percent
6. Measurement
7. Geometry
8. Solving Equations

Introductory Algebra: Concepts and Graphs

1. The Basics
2. Linear Equations and Inequalities
3. Graphing
4. Exponents and Polynomials
5. Factoring
6. Rational Expressions
7. Systems of Equations
8. Roots, Radicals, and More Quadratic Equations

Intermediate Algebra: Concepts and Graphs

1. Numbers, Variables, and Expressions
2. Equations and Inequalities in One Variable
3. Equations and Inequalities in Two Variables
4. Systems of Equations
5. Rational Expressions and Rational Functions
6. Rational Exponents and Roots
7. Quadratic Functions
8. Exponential and Logarithmic Functions
9. Sequences and Series
10. Conic Sections

Essentials of Elementary & Intermediate Algebra: A Combined Course

1. Linear Equations and Inequalities
2. Linear Equations and Inequalities in Two Variables
3. Functions and Function Notation
4. Systems of Linear Equations
5. Exponents and Polynomails
6. Factoring
7. Quadratic Equations
8. Rational Expressions and Rational Functions
9. Rational Exponents and Roots
10. Exponential and Logarithmic Functions